Quantum Relativity

from Cosmology Architecture to Chromodynamics

BLACK and WHITE Edition

© 2018 by
PůMa Tse

All rights reserved
Akademé (http://akademe.org)
ISBN 13: 978-1987666953
ISBN 10: 198766695X

Proceeds of Akademé materials and services support Akademé Foundation's mission to promote sustainable civilization through education and advancing understanding and compassion.

Cover Photo: MACS J0416.1-2403 at z=0.397
Credit: ESA/Hubble. (Mar. 21, 2016). A Cosmic Kaleidoscope. https://www.spacetelescope.org/images/potw1612a/.

We want to start by thanking our critics....

    We especially want to thank readers for feedback and students for their questions.

Contents

Figures/Equations IV
Introduction 1

Significance	1	Quantum Forces	4
Sourcing	2	Constructing Spacetime	4
Outline Summary 2		Quantum Morphology	4
A Bookshelf	3	Instantiation	4
Void Basics	3	Relative Field Theory	5
Creating Singularities	3	Confined Morphology	5
Change Functions	4	Degeneration	5

Cosmology 7
Cosmology Architecture   8

1. A Bookshelf 9

Perspective 9		Occam's Cycle 19	
Evidentiary Issue	11	Cyclic Functions	20
Architecture 13		Quantum Shade 21	
Möbiverse 15			
Value Redistribution	17		

2. Void Basics 25

Point into Dimension 25		Distance 32	
Geodesic Field Equation		Light Distance	33
(GFE)	26	Distance Functions	33
Sea Level 27		Comovement	36
Light & Photons	29	Into Shade 37	
Expanding Void 30			

3. Creating Singularities 39

Mathematical Root 39		Creating Singularities 46	
Quantum Approach 40		Helicity v. Chirality	48
Thermodynamics 43		QR Field Shapes 50	
		Redefining Entropy	51
		Vortices 53	

Quantum Field Theory 55

    Change-Color Truth Tables   56   Permittivity & Permeability
                                                                      Constants   56

4. Change Functions 57

Entropies	57	Algebraic Logic	63
Into Color	59	Polar Refresher	66
Venn Algebra	61	Context Graphs	68

5. Quantum Forces 71

Waves	71	Bose-Einstein	79
Quantizing	75	Adaptations	80
G & the Vector Axis Modifier	76	Intrinsic Mass	83

6. Constructing Spacetime 87

Background	87	Inactive Surfaces	95
Basic Interactions	91	Tensor Manifolds	96
Brane Surgery	92	Term Declarations	97
		Lie Groups	100

Matter 103

    Periodic Particle Table   104

7. Quantum Morphology 105

Generations of Matter	106	Gluons	112
Virtual Particles	107	Ambiguous Cases	113
Primordial (Color Charges)	109	Weak Confinement	113
Weyl Fermions (Flavor Volumes)	109	Leptons	114
Unbound Identities	111	Quarks	117
Photons	111	Hadronization	117

8. Instantiation 119

Strong Type I	119	Mitosis v. Meiosis	126
Cladomorphology	122	Interphase	130
Baryogenic Asymmetry	125	Eigenstates	132

9. Relative Field Theory 135

Quantum Gravity	135	The Electroweak Field	140
Renormalizing	136	MicroG & Fermi Surface	142
GFE Fields	138		

Degeneracy 147

10. Confined Morphology 149
Trionic Bonds 149
Structure Groups 151
   QM Aspects   153
Quantum Dynamo 154
Electron Orbits 157

11. Degeneration 161
What is Degeneracy? 161
Beyond Wonderland 165
General EM Fields 167
Degenerate Matter 172
Understanding the Matrix 173
Conclusion 176

Quantum Relativity

Figures/Equations

Front of section figures are in the table of contents.

1.1: ESA—Hubble Ultra-Deep Field z≈10 10
1.2: Galactic Life Cycle 11
1.3: Hubble's Tuning Fork 12
1.4: Cosmology Architecture 13
1.5: Redistribution, Chainsaw Example 14
1.6: Overlapping Bubbles 16
1.7: Value Redistribution Function 17
1.8: Quantum Shade Flowchart 18
1.9: Poisson's Gravity Function 20
1.10: Helical Time & Occam's Cycle 21
1.11: Spacetime Fabric and Quantum Shade 23
2.1: Geodesic Field Equation Elements 26
2.2: NASA/WMAP CMBR Anomaly 27
2.3: GRACE Gravitational Anomaly 28
2.4: Light Spectrum 31
2.5: Anchorage to Nome 33
2.6: Distance Elements & Calculations 34
2.7: Co-movement v. Expanding Void 36
2.8: Minkowski Space & Dilation Diagram 37
3.1: Tap Shape Permeates fill into Cup's Permitted Shape 41
3.2: Perkins & Rotating Heat into Cold 44
3.3: Fleming's Rules & Force Variables 45
3.4: Singularity/Brane Field Equation with Color Charge 47
3.5: Helicity 48
3.6: ESA/Hubble "Dark Matter" Kaleidoscope 52
3.7: Phase on Red/Cyan Singularities 53
4.1: Quaternions v. Tessarines 58
4.2: Boolean Logic Symbols 58
4.3: Color to Change/Charge & Palette 59
4.4: Color/Change Truth Tables 60
4.5: Color/Change Operation Diagrams 61
4.6: Evolving Venn Diagrams (to Hypercomplex) 62
4.7: Change Logic Rules Table 63
4.8: Quadratic j-Change Example 64
4.9: Imaginary Foils/Roots 64
4.10: Hypercomplex Phase Logic 65
4.11: Quadratic i-j Relationship Example 65
4.12: How to Polar Plot 66
4.13: Circular Function Polar Plotting 67
4.14: Tan/Cot Polar Plots 67

4.15: 3-D Algebraic Context Graphs of Changes..68
4.16: Flat Radian to Cross-Section Profile..69
4.17: Applied Circular Functions to j&i Changes..69
4.18: j&i Change Planes 70

5.1: Fleming's Rules & Force Variables..71
5.2: Scalars in Quadratic Equation Example 72
5.3: Classical Force, Spin, Frequency, Momentum 72
5.4: Longitudinal, Transverse, Surface Waves 73
5.5: Polarized/Unpolarized Waves..73
5.6: Filtering Polarization 74
5.7: Optical Bandwidth 74
5.8: Computing Newton's Constant (G) 76
5.9: Triangulating Force in Color Distribution...77
5.10: Bose-Einstein Microstate Example 80
5.11: S-entropy as Energy Density 81
5.12: Microstates of Like/Unlike Binary Virtual Particles........................81
5.13: Microstates of Singlet & Topolariton Virtual Particles 82
5.14: Affect of Microstates on Intrinsic Motion 82
5.15: Possible Microstates of Weak Bosons...83
5.16: Possible Microstates of Quarks 84
5.17: Possible Microstates of Nucleons 84

6.1: Complex Functions & Axes in Sphere 87
6.2: Axial Rotations of Complex Change Variables 88
6.3: Haar Linear Axis Relationships to Cube (4-D) 89
6.4: Interactions Compared 91
6.5: Rotating Aspects Example...92
6.6: j-Right i-Left Branes 92
6.7: Elliptical Functions 93
6.8: NASA/WMAP CMBR Elliptical Image (#121238).............................93
6.9: Toroidal Functions 93
6.10: Toroids Rotating into Phase-Spheres 94
6.11: Lorentz Dilation Function & Graph...95
6.12: Band/Flux Tube Synchronicity (Chi Twisting) 95
6.13: Evolution of Variables Matrix 97
6.14: Periodic Matrix 98
6.15: Lie Algebra Basics 100
6.16: Lie Elements of Sub-manifolds 100
6.17: Right-Left Lie Fields 101
6.18: Lie Derivative Definition 101

7.1: Hierarchy of Life v. Generations of Matter 106
7.2: Virtual Particle Quantum Constructs..108
7.3: Confined Color to Weak Charge Equivalence 112
7.4: Confining Particles Pre-Structural Compositions 115
7.5: Michael Evan's Trion and Trionic Bands......................................116
7.6: IQ Puzzle Ball/Quark Bonding Example 117
7.7: Trionic Band Synchronicity 118

Quantum Relativity

8.1: Complex Toroidal-Phase-Interactive Singularity Fields 121
8.2: Cladomorphology Types and Thermodynamics 122
8.3: Quark Decay Path with Flavors Indicated 126
8.4: Band Growth to Mitotic Snap 127
8.5: e to q + g Anangenetic Diagrams 127
8.6: d to u + e + v Anagenetic Process 128
8.7: Feynmann Diagram of d to u + e + v 129
8.8: Enthalpy Process Diagram of d to u + e + v 129
8.9: Interphase Process 130
8.10: Schrödinger Normalization of x in Wave (ψ) over time (t) 131
9.1: Einstein's Brane Field Equation 136
9.2: Value Redistribution Function 137
9.3: GFE Boundaries and Conversion to Momentum 139
9.4: Electroweak Field Explains Cabibbo Matrix 141
9.5: GRACE Satellites Ride g-Waves 143
9.6: GRACE's Mantle Gravity Anomalies 144
10.1: Trionic Band Edge-Surfaces and Bonds 150
10.2: Regular Solids Give Structural Rules 151
10.3: Ferromagnetic Nuclear Shells to Field Lines 155
10.4: Regular Solid Features 156
10.5: Mensuration of Regular Solids 157
10.6: Regular Solid Ex-Inscribed Radial Ratios 157
10.7: Wiswesser Notation & Equation 158
10.8: Electron Orbit Eigenstates 158
10.9: Electron Orbit Paths in Quantum Solids 159
10.10: Periodic Table Limitations 160
11.1: Classical Navigation Axes and Angles 163
11.2: Navigation by Quaternion/Change Axes 164
11.3: Saturn's North Pole 166
11.4: Electroweak Field Equations 167
11.5: Magnetic B Vector and Free Energy Fields 168
11.6: Conic Section Recti-Polar Definitions 170
11.7: Basic Elliptical Orbit 171
11.8: Strong Type VII—Foam Redefinition 172
11.9: Periodic Matrix Surface Features 174
11.10: $E=mc^2$ Matrice Form 175

Introduction

> The more success the quantum theory has, the sillier it looks.
>
> —Einstein, May 20, 1912
> to Heinrich Zangger, AEA 39-355

This book is primarily about constructing, interacting, and evolving spacetimes. This means understanding how change functions define quantum fields, color charge, morphological patterns of matter, fields, and degeneracy. We set aside all expectations and theories, take a pure mathematical perspective, dissect and ruthlessly apply the rules all the time. Let us reconstruct how the universe does things from the bottom up.

Now we can ask where did the Big Bang come from? Where will it go? How does it all work? How do big bangs evolve? Does the universe of big bangs also evolve? How does time work? The expansion of space? Why does light propagate as it does? How do singularities work? How are space, matter, energy and light created, interact and evolve?

How do we get our observations to agree with the theories explaining how they work? How do we establish a working cosmology that follows all the rules all the time? How do we make it accessible and socially useful? We even stand a chance of observing and answering without logical fallacies or losing falsifiability.

Less than 5% of the universe is conventional matter. Why are we so convinced we can explain the universe from our perspective as an nth-degree anomaly? The proportion of dark matter (ν) to dark energy (μ including conventional matter) is $1:e$. Let's approach and answer our questions from the universe's perspective, from their angle. Let us apply all the rules all the time instead of throwing them out at our convenience.

Significance

Cosmology is the structure we use to put all our scientific investigations in a working order and find our own place in the universe. It is vital to scientific integrity and the sustainability of society. It is not meant to be known like religion. Science understands by enabling diverse expertise and putting hard evidence ahead of theory.

Cosmology is a quantum complex of contexts and nuances. It does not fit in any matchbox, children's story, or even the deepest line of theory. We are not here to bash or throw out any working theory including Big Bang Theory. All these must work together in an architecture if treated honestly or fail alone.

The architecture here depends heavily on the workings of Thermodynamics, quantum field theory, and chromodynamics (QCD)—the

Quantum Relativity

Standard Model of strong particle interactions since 1972.[1] This dependence in no way is an exclusion or to give rank. Each thing has its place. These things just happen to stitch others together in workable forms.

To solve evolved problems we need evolved thinking, language, and concepts. While we purposefully slow and cite the backgrounds going into the thinking, it is vital to pay close attention to our definitions, and take seriously words we fairly pillaged from the creative human universe for their concepts. We are explicit with evolving cause. Never assume you already understand.

Our biggest innovation isn't the cosmology. The cosmology isn't as original as some may think either. The innovation is how we define change functions. Also not exactly new, just that we are explicit and show how those complex variable definitions shape things. Then we apply them and show how different contexts form a logical sequence of evolving material morphology. This helps see what confinement and renormalization hide, how degeneracy works and eventually evolves into singularities.

Sourcing

Our sourcing choices are primarily pedagogical, but also authoritative. We aren't just randomly providing a long list of sources agreeing with any one thing. Instead, we are going to the leading expert those sources used for vital things or for educational value.

If it is something basic covered in a university class, we point you there. Otherwise, if you want to understand how things work, talk to the people getting their hands dirty, not the people in the other room sitting around a table with nth-hand information or what is mainstream or popular.

Science isn't popular, it is hard. We care what NASA, ESA, CERN, university laboratory research scientists and engineers addressing the observations and problems directly say. If you want to understand distance, for example, you go to the guy whose calculator is what everyone uses.

Outline Summary

An architecture is a bookshelf. It is what you put on its shelves. This book builds a bookshelf and then explores the protocols and logic leading to that bookshelf. These protocols dig into the roots of quantum field theory, how chromodynamics and spacetime work, the evolution of matter and emergence of complex fields from degeneracy. It is divided into four sections pedagogically arranged:

---

[1] Fritzsch, H. (Sep. 27, 2012). <u>The History of QCD</u>. cerncourier.com/cws/article/cern/50796.

Quantum Relativity

Cosmology (introduction)—paints the big picture and key concepts that make it work.

Quantum Field Theory (theoretical framework)—provides the mathematical logic to define color charges, how they are valued, the definitions and development of spaces into structure.

Matter (methodology)—shows how matter is created, defines its fields, interacts, and evolves from virtual through confining into degeneration.

Degeneracy (observations and conclusions)—features of nuclide degeneracy, conventional properties and interactions, magnetism, orbits and other EM. Closes the circle by normalizing degenerate information into singularity.

A Bookshelf

We step out of the Big Bang matchbox into the fire with a narrative outlining an even bigger picture within which Big Bang bubbles emerged. An OSI-like architecture is introduced to help organize established understandings into a comprehensive working theory. We then begin to examine the deepest concepts of this system so the reader has a general idea of its workings, and how the Big Bang fits in.

Void Basics

A long chapter on literal nothing. The universe uses points in spacetime to spread disorder and cause all work from the lowest level up. We examine the implications of CMBR on galaxies, and how this muddies our ability to know in the deep field. Distance functions are examined, explained, and shown how they are applied to give us a range of useful information.

We observe redshift $z=1$ is practically next to $z=cs/m$—where wavelength flat-lines. The reason is resistance to void expansion and the inverse wavelength-frequency relationship defining spacetime, not dilation or Minkowski. We also observe the adaptation factors are applied twice in the lookback distance function designed to squeeze the universe into Lemaître's matchbox.

Creating Singularities

Where Einstein's geodesics become indistinguishable, a juxtaposition occurs. One manifold enfolds the other manifold, which unfolds to all points of void in real time (phase). That value is CMBR. The fields are classified chromodynamically and evaluated for a broad range of interactions including Thermodynamics and strong interactions responsible for time, CMBR, evolving matter, and galactic processes.

Quantum Relativity

Change Functions

Tessarines, quaternions, and Boolean logic are explored to unveil a system of change as imaginary, complex, and hypercomplex functions. The concepts are then illustrated in a variety of contexts to clarify how the logic works algebraically and graphically. QCD color association and truth reveal the process of virtual emerging through interaction into real (relativistic).

Quantum Forces

Scalar energies at the root of all valuation derive easily from Fleming's rules. They transfer in wave form mechanically and electromagnetically. Quantization creates new virtual matter that must confine in interaction or annihilate. In interaction, the spaces are sequentially created and shaped as value of microstates flows through. The mass problem is easily seen as an array of elements contributing in degrees.

Constructing Spacetime

Euler provides the foundation to evolve change functions as linear values into surfaces, interactions into volumes, and establish the Laplacian. All apply simultaneously in different contexts. Change function axes provide sequence, direction, and shape to points. We then use Lie concepts to group, manipulate, and derive basic interactive field definitions.

Quantum Morphology

The hierarchy of life provides a similar sequence for matter. The evolution of seven strong interaction types follows this plot through to hadronization and nucleon interactions to form isotopes. The particle plot simplifies into a three act play: virtual particles, unbound identities, and weak confinement. We put Weyl fermions and Fermi surfaces to good use winding the maze of complex interactions to understand weak interactions, mass, and much more.

Instantiation

New matter creation is a potential of the first type of strong interaction. Again we must turn to biology for terms like mitosis, meiosis, and evolutionary processes not accounted for in physics. Inadequate language makes finding information in physics so difficult that even experts get lost and confused, wasting time with wrong questions like baryonic asymmetry.

Schrödinger and eigenfunctions help normalize, oscillate, and ultimately confine the details making matter do what it does.

Relative Field Theory

Relative fields start evolving with the initial quantum gravity brane. Einstein's brane curves spacetime into the geodesic definition and shows multiple paths to singularity. His geodesic equation is two branes interacting to form a third, which is the spatial action of gravity.

We refine the GFE to show its boundaries, conversion through Poisson-Gauss into momentum and acceleration. This reveals ambiguities like the electroweak field, microgravity, and the continuation generally without limits. Recognizing the context and definition of the electroweak field shows how CP-violation applies, how microgravity emerges, and how GRACE maps gravitational anomaly.

Confined Morphology

Atoms are the final stage of renormalization. Nucleons, electrons, and their interactions still have quantum qualities being confined through renormalization into atoms. Nuclide structure is confined in a permittivity (Schwarzschild radius) with a degree of disorder providing an imaginary space of interactions.

From this imaginary manifold applied to a fixed density, intrinsic magnetic qualities and induction emerge. It also means much of our theories about nuclide structure are simultaneously true depending on context. A similar examination of electrons reveals a layer of confined complexity that can explain valence states, limits on the periodic table, and likely many more otherwise inexplicable concepts.

Degeneration

Renormalization confines quantum details in degrees. Degeneration is a process of re-establishing quantum specifics by quantizing in degrees. This is where we emerge from too small to observe into observable and beyond our capacity to reasonably observe again. First is density, quantized at permittivity leaving a permeability quantum "foam." From this impurity emerge the first degenerate qualities like magnetism.

Second is incomprehensible numbers of individuals; so many together that despite differences, they form quantum mechanical behaviors and qualities. Third is the last type of strong interaction: the pursuit of change equilibrium (e.g. $S=0$). Around these, a myriad of electromagnetic forms emerge—many as yet to be explored. The Matrix and field equation forms help to some degree to analyze and give us ideas to explore further.

Quantum Relativity

Cosmology

We are all agreed that your theory is crazy. The question that divides us is whether it is crazy enough to have a chance of being correct.

—Niels Bohr, 1958
to Wolfgang Pauli re: Heisenberg Uncertainty[1]

---

[1] Dyson, F.J. (Sept. 1958) "Innovation in Physics." *Scientific American*. 199, No. 3, pp. 74-82. scientificamerican.com/magazine/sa/1958/09-01/.

Quantum Relativity

Cosmology Architecture

USE Mode				PHASE Mode					
Classical — Mechanics, Momentum	**Relativity** — Spacetime, Dilation	**Strong Interactions** — Subspaces, Color Charges		**Expanding Void**	Celestial WIMPs	Quantum Shade — holistic interconnectivity	Occam's Cycle — change simple→disorder	Möbiverse — the superposition twist
					Interaction			
					Information			
					Neutron Stars / Pulsars			
					CMBR			
					SM Black Holes / Quasars			
				Thermodynamics				
Optics	Mass	Light						
Work	Time	QCD	Unfolding Universe	Enfolding Multiverse				Abstract Phase

PŭMa Tse—8

1. A Bookshelf

Once upon a time long long long ago.... Perhaps even longer ago....
A very special kind of black hole formed by fully occupying its surface and expelling its volume in one giant burst of light focusing around its equator and forming the first particles of the Milky Way (as a Lyman-alpha emitting galaxy).

Gravity contracted space around the poles of the black hole, drawing in the newly formed matter to form a bar-shaped spiral. It sucked all this matter and more slowly back into its volume. Some strongly interacted and became part of the black hole.

When something is lost forever to the black hole and similar objects, its value gets distributed to the entire universe as CMBR and expanding void. This is how gravity, strong interactions, and entanglement keep the universe and time going.

Every so often, the volume of the black hole fills up enough to cause another burst. Each burst contains light forming new matter like protons and electrons, and old matter evolving into massive WIMPs and heavy elements.

One such burst happened about 13.7 billion years ago, shaping the Milky Way and eventually our solar system and world as we know them today.

Perspective

A vivid story creates an idea space within which you can build a system. Such a system that organizes established working models of physics is a cosmology. Science evolves, so the ideal system is an adaptive architecture. If something doesn't fit the evidence, you replace it with what does. Instead of cosmology defining the parts, the parts define the cosmology.

Big Bang Theory (BBT) is one of our most popular stories. It isn't perfect, but it has an important place. It is a simple and rigid storyline lacking evolution of itself. Created things evolve, come from, and go. The rigidity has come at odds with observational astronomy and quantum physicists the world over.[1] Astronomers have become outright cynical about the conflicts between BBT and observations.

Dr. Wright is stuck at a desk in the UCLA Astronomy Department. He developed the calculator everyone uses to compute distance. Within minutes of a press release, he gets questions like, "If a distant cluster of

---

[1] NASA. (Apr. 16, 2010). Beyond Big Bang Cosmology. map.gsfc.nasa.gov/universe/bb_cosmo.html.

galaxies is 9.1 billion light years away in a universe that is 13.7 billion years old, how did the cluster get so far away in only 4.6 billion years?"[2]

1.1: ESA—Hubble Ultra-Deep Field $z \approx 10$

    Hubble's Ultra-Deep Field (HUDF)[3] captured 10,000 galaxies in the constellation Fornax. This is a full range of galaxy types and levels of development dating back 400 to 800 million years after the Big Bang. OUR Big Bang.[4] The universe is more complex and bigger than Lemaître ever conceived.

    Science is about understanding, not knowing, and certainly not popularity or appearances. Lemaître, as a priest and source for BBT, in a way opened up a mindset of thinking science is like religion: about knowing. This is unhealthy. As Dr. Wright laments, we know exactly what we observe, which is the redshift information. Beyond that we enter the realms of speculation and hypothesis.

---

[2] Wright, E.L. (2013). Why the Light Travel Time Distance should not be used in Press Releases. astro.ucla.edu/~wright/Dltt_is_Dumb.html.
[3] (Dec. 8, 2009). Hubble's Deepest View of Universe Unveils Never-Before-Seen Galaxies. hubblesite.org/image/2644/news_release/2009-31.
[4] Bouwens, R.J. et al. (Jan. 27, 2011). A candidate redshift $z \approx 10$ galaxy and rapid changes in that population at an age of 500 Myr. https://www.nature.com/articles/nature09717?page=5.

Quantum Relativity

The real problem of BBT is the elephant in a matchbox.[5] Unlike the famous legal case, science has been trying to squeeze the elephant into the matchbox. The matchbox is itself fine. Like Darwin's version of evolution, it may have some pre-adolescent awkwardness here or there, but generally speaking, the theory actually works and checks out.

Evidentiary Issue

Lyman-alpha radiation (LAR) comes from hydrogen creation.[6] CMBR allegedly is Big Bang "afterglow." No LAR has been observed in CMBR.[7] LAR is observed in our solar system and galaxy, so hydrogen is not just created one way.[8] The most efficient observed way to mass produce large quantities is in an LAE (emitting) galaxy.

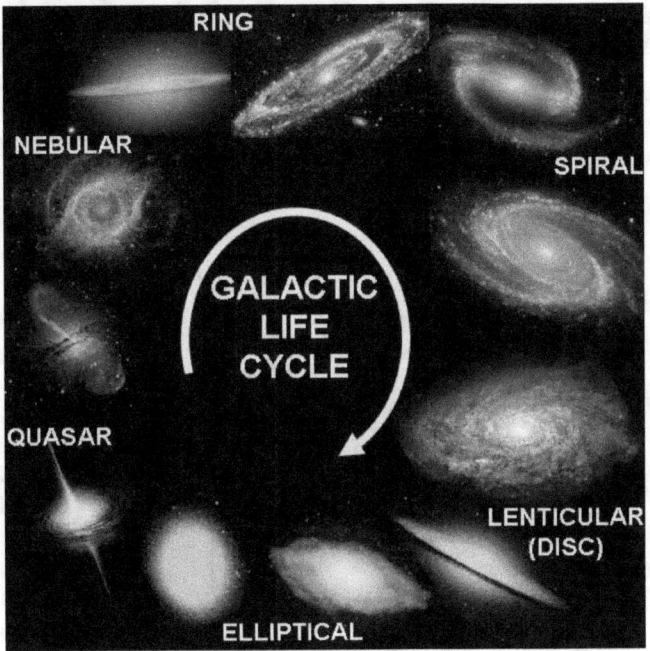

1.2: Galactic Life Cycle

---

[5] Mapp v. Ohio, 367 U.S. 643 (1961).
[6] Draine, B.T. (2010). Physics of the Interstellar and Intergalactic Medium. Princeton, N.J.: Princeton University Press.
[7] Brown, M. (ca 2014). Cosmic Microwave Background. JodrellBank Centre for Astrophysics, Manchester University. jodrellbank.manchester.ac.uk/research/research-groups/cosmology/cosmic-microwave-background/.
[8] Castelvecchi, D. (Dec. 1, 2011). "Voyager Probes Detect "invisible" Milky Way Glow." National Geographic. news.nationalgeographic.com/news/2011/12/111201-voyager-probes-milky-way-light-hydrogen-sun-nasa-space/.

Quantum Relativity

LAE is a type of early stage galactic development. Field dynamics show this as an early stage of galaxy-singularity development. That function is a massive burst of energy focusing enough to forge a particle zoo. Each successive bang evolves the nature of what is processed. The second bang will contain degenerate matter where the first was energy-based.

From this zoo, protons and electrons naturally emerge and conjoin among other particles like neutrinos. When LAE's start forming stars they are called Lyman-break galaxies. Similar later bursts, like Lemaître's Big Bang, will produce rare Earth elements. His was definitely not our galaxy's first big bang. It was an evolved bang.

The galactic cycle illustrated takes billions if not tens of billions of years depending on the scale. It is just the cycle and doesn't illustrate that the cycle evolves each time. Hubble was first to show this process beginning with his 1926 classifications.[9] Until 1936, however, he had not yet established that "nebulae" were separate from our galaxy. This is why we set Big Bang aside to follow the evidence and find its place.

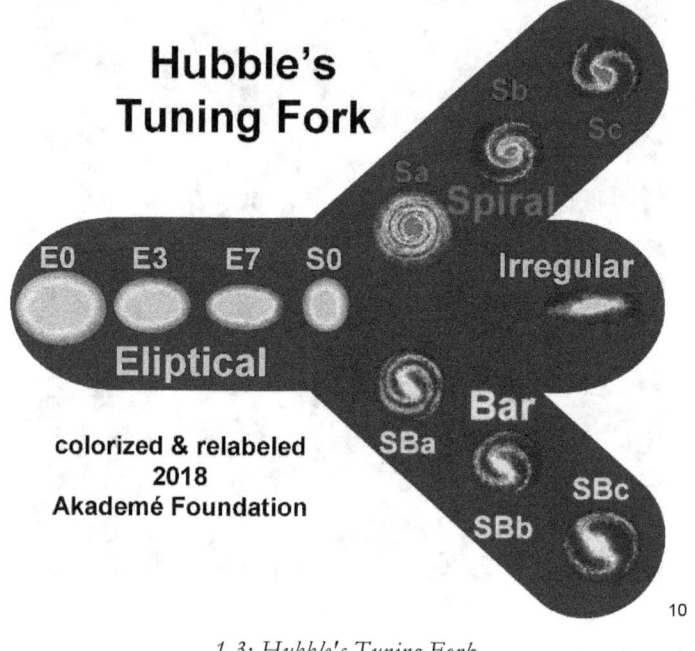

1.3: Hubble's Tuning Fork

---

[9] Hubble, E.P. (1927). Extra-galactic nebulae. Astrophysical Journal. aldebaran.cz/astrofyzika/struktury/galaxie/docs/Hubble-1926ApJ_64_321H.pdf.

[10] Koberlein, B. (Apr. 25, 2014). "Hubble's Tuning Fork" in Galaxies. https://briankoberlein.com/2014/04/25/hubbles-tuning-fork/.

Architecture

We do not live in a classical or relative universe. We live in a highly sophisticated quantum universe from which relative and classical things emerge. Like a computer system, the actual core components and processes are incomprehensibly simple, hyper-connected, and numerous.

We are the end users of this system, skewing our view of its nature. Like every end user, we think what we see is what it is. Then we use a hammer to break open the box only to find the computer inside looks nothing like what we see in use. This makes physics really hard.

QCD Time Work			Classical			USE Mode
		Optics	Mechanics		Momentum	
			Relativity			
		Mass	Spacetime		Dilation	
			Strong Interactions			
		Light	Subspaces		Color Charges	
Unfolding Universe	Thermodynamics		**Expanding Void**			PHASE Mode
Enfolding Multiverse		SM Black Holes Quasars	CMBR	Neutron Stars Pulsars	Information Interaction	Celestial WIMPs
Abstract Phase			Quantum Shade—holistic interconnectivity			
			Occam's Cycle—change simple→disorder			
			Möbiverse—the superposition twist			

1.4: Cosmology Architecture

On top of the physical system is a program-like architecture making it usable. Our first and most obvious architectural distinction is between our end-user perspective and the physical system. We will call the end-user portion our USE Mode, and the physical system our PHASE Mode.

Phase has many definitions in physics. One is "the position of a point in time (instant) on a waveform cycle."[11] Phase describes a moment of unspecified duration and all the change conditions fitting into that moment. Time never occurs in PHASE Mode Time only applies conditionally in USE Mode. "Expanding void" isn't expanding until USE Mode applies.

Time is resistance to the change interaction defining it—a component concept group of the USE Mode. Time is used to increment the effects of change into a sequence and as an effect increment (unit). Time derives from strong interactions in phase, such as interactions of entanglements as

---

[11] Rouse, M. (Sept. 2005). Phase. whatis.techt2arget.com/definition/phase.

Quantum Relativity

surfaces interacting with volumes of strong bonds.[12] There are many types of strong interaction we will cover in this book.

This surface-volume interaction specifically satisfies the geodesic field equation (GFE) defining mass and surface gravity. This emergence of time is an effect increment. It resists change in mass and acceleration—conventional dilation. This is time emerging from USE Mode variables. Time emerging from PHASE Mode gives us the familiar concept of a sequence of events.

Starting at the bottom, the "kernel" is the OSI (Open Systems Interconnect) term for the PHASE Mode.[13] It is the part of an operating system in complete control. The system itself is technically **inert** without the USE Mode. That has nothing to do with the presence or absence of human observers. We just happen to see things from USE Mode perspective.

To understand PHASE Mode objects, we often put them in a USE Mode context. Each PHASE piece is merely a function set to be called into USE application contextually. In programming these would be modules—algorithm units in an object-oriented program called into action as needed. Thermodynamics acts as the main controller determining which parts apply and in what order to include simultaneous.

At the root are the three Abstract Phase elements. As the term suggests, they are hidden concepts like superposition, cycle, transformation, interconnectivity, and conservation. We give them catchy names to help put them into their own contexts.

Each is incredibly practical. Consider the chainsaw below. The cycle is only possible thanks to the "twists" in the system giving the chain direction. The system itself is an interconnected thing. A change in one place affects everywhere. In particular, the enfolding/unfolding are simultaneous, conserving by transforming. Not only are variables transforming, they and the transformation are doing other things at the same time.

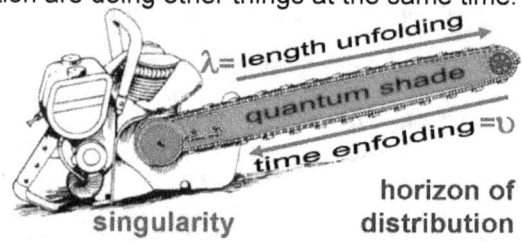

1.5: Redistribution, Chainsaw Example

We prefer the term singularity to black hole. It is more meaningful and mathematically relevant. A singularity enfolds a surface in perfect order $S=0$. For this to work under the laws of Thermodynamics, ultimate disorder

---

[12] Wolchover, N. (Apr. 16, 2014). <u>Time's Arrow Traced to Quantum Source</u>. https://www.quantamagazine.org/quantum-entanglement-drives-the-arrow-of-time-scientists-say-20140416/

[13] Yool, G.R. (2014). <u>Practical Algorithms</u>. 3 ed. pg 55.

must apply somewhere else in the system. That must happen at the greatest possible distance, which is the light horizon. The encapsulation we will call a Möbiverse **bubble**.

The unfolding location at the generic light horizon is $ch^{-2/3}$ seconds condensed by actual horizons and bubble interactions to <46.85 Gly. This allows for distribution of the full spectrum. The actual horizon is proportional to the magnitude of the enfolding surface unfolding as wavelength to enfolding cycle segment (u) by:

$$\upsilon = u/[1.3E22 \text{ seconds} = (Uh)^{-2/3} \text{ seconds}]$$

U is unit conversion. The spectrum agrees with this limit by dead ending at maximum dilation shortly after E24 seconds (see pg. 31). The horizon distance is specified by the maximum wavelength $\lambda = c/\upsilon$.

The change function defining this bubble is itself outside of time. It is the conservation and holistic interconnectivity. We named it Quantum Shade[14] because it defies all conventional notions of time and space to maintain exact proportions. This is often done in obvious ways like converting wavelength to frequency. These are distinct inter-dependent things, one expanding while the other contracts. At other times, as with annihilation, it is harder to see because everything is transformed radically and not necessarily locally.

Put into application, linear spacetime $c=\lambda\upsilon$ generically spins out no matter what the parent variables are doing. The parent variables do affect dilation local to the galaxy and singularity, and exponential wavelength growth toward the horizon (Doppler effect). This spacetime is not just an effect setting all our constants—and constants aren't just numbers, they do things. It expands every point in space at c resisted by local change conditions. This expansion gives us light and makes matter work.

Möbiverse

The Möbiverse is superposition with a twist. Wherever you are is the center of the universe. More accurately, the center is the nearest singularity. The ideal light horizon emerges from the interactions of actual horizons as the ultimate twist point. Your whole universe is your bubble plus the part of all intersecting bubbles to some degree or other.

Linde's bubbles are inflationary.[15] These inflate with growth of the enfolding manifold but generally experience density fluctuations. It is highly likely that a singularity twists itself in a big bang to create such a bubble. Even if that isn't the case, the bubble exists because of the singularity, and will persist as a density distortion after the singularity is gone.

These bubbles are the overlapping quantum environments of **Big Bangs**. They do not come together in a neat orderly manner to define a

---

[14] Dollard S. (Nov. 21, 2015). "Face the Raven." Dr. Who. BBC.
[15] Ferguson, K. (2012). Stephen Hawking: an Unfettered Mind. Macmillan.

consistent shape. Instead they form interactive patterns. At their cores is or was a singularity.

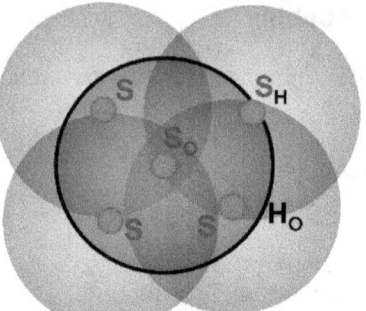

Edge horizons = greatest expansion/unfolding
Observer's
S_O=Singularity/Enfolding
H_O=Light Horizon
S_H=S near H
Blue density=CMB flux

1.6: Overlapping Bubbles

A singularity is an inert virtual identity of ideal order. Virtual means it is a perturbation dependent on its environment and other conditions, namely disorder. Without disorder, a singularity becomes active long enough to enfold out of existence. The horizon is the ultimate disorder, but it is the local disorder that maintains singularity, preventing it from enfolding into its annihilation.

Mathematical singularity is often equated with degeneration[16] from normal behavior and spatial definition. The environment bends space into the singularity creating a massive gravitational effect. Singularity occurs when the interacting geodesic manifolds become indistinguishable.

These geodesic manifolds are mathematical abstractions known as branes—surfaces without intrinsic shape or depth. Each has its own path to filling a quantum number. When one fills its quantum number, it becomes indistinguishable and can trigger singularity or exclusion.

The second space becomes available to be filled with mass pushed into it by the environment. Typically it starts with value in it. A big bang (quasar, pulsar, etc.) happens when this second space reaches its quantum number. While mass accumulates, it is subjected to information equilibrium (see Wheeler interaction pg. 172 et seq.).

Information is the absorption pattern of order in light (EMA), and light gives value to the systems of matter. The radiant part of light (EMR) is the distributive disorder. The pattern directly affects how the change functions

---

[16] Weissstein, E.W. (2018). Singularity. From MathWorld--A Wolfram Web Resource. http://mathworld.wolfram.com/Singularity.html

Quantum Relativity

of matter define the spaces. It is a requisite level of disorder specific to a material identity.

The level of disorder for a singularity is S=0. Anything achieving that optimum order adds to the singularity effect. The change in value translates into CMBR at the light horizon. It also increases the ground state value of the singularity. When a singularity gains mass or enfolds it grows. When it expels it bounces back to ground state scale.

These overlap as the picture shows, so the CMBR effect observed here is from positions near our horizon in the present. This naturally affects the motions and positions of galaxies. It can take up to 46.85Gly for CMBR emission to get back to us, but it would be impossibly weak. What we are observing is comparably strong, but the effect is more obvious on large scales. And it is not the evolving bursts we would consider big bangs.

Value Redistribution

Quantum Shade provides interconnective conservation. Conservation transforms proportionally and oppositely like Newton's 3rd law.[17] A singularity enfolds value (ultimate confined order ν=nu) by converting the information in matter into unfolding wavelength at its actual light horizon. This generates the process. The actual horizon (see pg. 15) is the distribution point of ultimate unfolding disorder (μ=mu).[18]

$$\upsilon_\gamma = \frac{1-\eta}{h} \left[\mu_{n-1} = \frac{m_{n-1}^{\,2}}{\nu_n} \right] \left(\frac{\nabla^2}{t^2} = c^2 \right)$$

mu at least = lesser m² in Laplacian spacetime units
upper magnitude singularity

$$\text{thermodynamic efficiency} \quad \eta = \frac{\delta W = PdV}{\delta Q = TdS} \quad \begin{array}{l}\text{Ejected value}\\ \text{Enfolded value}\end{array}$$

1.7: Value Redistribution Function

The unfolded bit of information is a cycle. Going toward the singularity, time increments the cycle and dilates. The cycle as light flux (flow) is CMBR: frequency (υ=mc²/h) applied to a cycle length (λ=c/υ) as linear unfolding spacetime.

Dividing by Plank's constant converts energy to frequency by means of spin: a light distribution at the actual horizon. As the Laplacian suggests, the distribution maps on a surface. The opposite natures of wavelength and frequency puts distribution relative to the light horizon per singularity (see pg. 33 et seq.). Distribution is simultaneous, the effect is spacetime with exaggerated spatial stretching toward the horizon to the rate of c.

---

[17] Henderson, T. (2018) Newton's Third Law. physicsclassroom.com/class/newtlaws/Lesson-4/Newton-s-Third-Law.
[18] Talbot, M. (1996). The Holographic Universe. London: Harper Collins.

Quantum Relativity

A singularity by itself cannot draw in light, but it can conditionally draw in masses trapping their light. Each singularity (v=nu) draws in value (Q) either through its poles or equator, then discharges (W) value through its equator (in bursts) or its poles (in streams). These discharges range from light to matter, both come into focus creating or evolving new matter.

Efficiency of discharge is a percent ratio η=W/Q. What is lost to Wheeler interaction is to us the inefficiency 1–η. This 1–η part of the value of m put into v changes the local radial values of the singularity. Their change affects the magnitude of the local spacetime.

Distribution is represented by the Laplacian. A Laplacian operator is used to represent a subjective coordinate system—meaning you plug in the one that works for you. It also allows for anomaly: uneven distribution of values.[19] The flowchart uses concepts throughout the book.

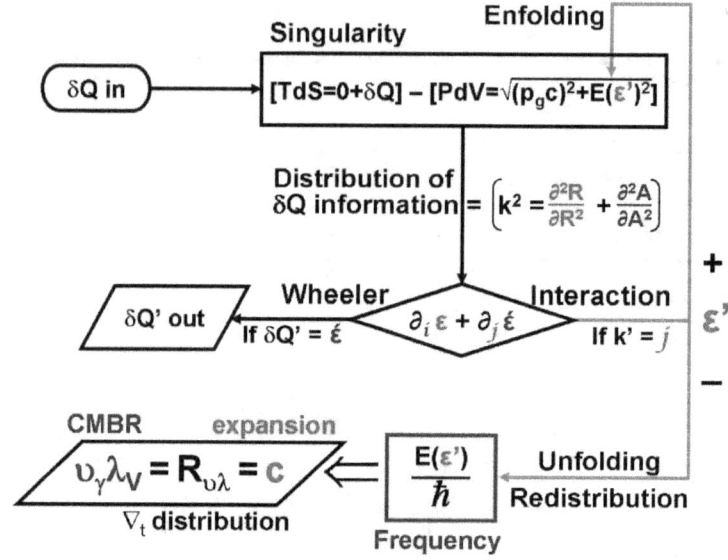

1.8: Quantum Shade Flowchart

These concepts convert to chromodynamic fields: singularity=v (red and cyan) and disorder=μ (blue and yellow). This is significant because μ is shaped by the sinusoidal change function i. It is a generalized virtual identity for all void points as a distribution field. Some may think of it as a virtual aether, the change in all points affecting everything not enfolded.

---

[19] Grinfeld, P. (Feb. 12, 2017). <u>What is the Laplacian?</u> Philadelphia, PA: Drexel University & Lemma. youtube.com/watch?v= 4J74tquQ7jU.

Occam's Cycle

Change cycles are vital to understanding the quantum universe on all levels. This specific component of the architecture contains the handling component of the change functions we would see as cyclic. This means that as value is applied to a change component, it unfolds in a specific order. This gives pattern to information process. The change functions and their handling of information defining spacetime is the entire Field Theory section.

Occam's razor states that given two solutions with the same outcomes, the simpler is preferred.[20] If both are valid, then this principle should not exclude the alternatives. It is also used in argument: the reliability of a premise is inversely related to the number of assumptions. The more assumptions made, the more convoluted and less reliable the conclusion.[21]

We use Occam to describe the simplest possible path to solution. The universe satisfies this in many ways to include:

- Multifunctionality—Function is defined by context and multiple functions that can apply simultaneously. This also leads to indifference among quantum states. Partial Differential Equations is a system of unknown change variables.[22] PDE multifunctionality applies to variables and functions. Interpretations evolve with understanding.
- Putting Nothing to Use—The universe uses nothing in the literal sense of nothing being a thing, like an absent part of a proportion, absorption lines on the spectrum, void as the difference between a whole and its parts, etc.[23]
- Forcing Variables—Squeezing things into spaces they otherwise wouldn't fit in, like variable spacetime, treating a letter as a number, or treating light as a virtual photon or in interaction as real.
- Path of Least Resistance—e.g. from high to low potential, but does not exclude other paths.[24] Everything functions by its definitions in such a way that nothing is ever actually consumed (Noether's theorem).[25] The universe essentially does nothing in the most spectacular and ambiguous way.
- Cycle as Identity—"The wave function for a given physical system

---

[20] "Entities are not to be multiplied without necessity," J.Poncius (1639) in: Scotus, J.D. (1894 reprint) Opera Omnia, vol.15, p. 433a. Paris: Vives.
[21] Gibbs, P. & Hiroshi, S. (1996-97). What is Occam's Razor? University of CA Riverside. http://math.ucr.edu/home/baez/physics/General/occam.html.
[22] Grigoryan, V. (Dec. 2010). Partial Differential Equations. UC Santa Barbara. p 1. http://web.math.ucsb.edu/~grigoryan/124A.pdf.
[23] Weatherall, J.O. (2016). Void: The Strange Physics of Nothing. Yale University.
[24] Holt, M. (Jul. 1, 2001). The Path of Least Resistance. EC&M. Overland Park: KS. http://www.ecmweb.com/content/path-least-resistance
[25] Noether, E. (1918). Invariant Variation Problems. Nachr. D. König. Gesellsch. D. Wiss. Zu Göttingen, Math-phys. Klasse. pp 235–257.

Quantum Relativity

contains the measurable information about the system."[26] A cycle permutation thus contains all the elements of definition for an identity. Identity, however, does not mean individuality.

The Occam's Cycle of the universe is the simplest means to establish all its parts and definition. The universe is a self-contained everything that wastes nothing. It is the perfect balance of yin-yang, intrinsic-extrinsic, absorption-emission, contraction-expansion. Time is uniquely defined relative to the context of its definition. There is no universal Relativistic time. The cycle, however, has its own clock—proper time described earlier.

Occam's Cycle time resists transformation between multiplicity and singularity. It is universal because each bubble is a self-contained proportion, their densities irrelevant to the whole. Applied with the superpositional features of Abstract Phase, this temporal definition theoretically limits the universe. It can also mean endless universe. Either hypothesis is equally valid and at this point, unprovable. You are definitely the center of your universe looking back to the history within your bubble.

Cyclic Functions

These concepts boil down to the initial field conditions of two color charge classes of matter in QR: red (ν=nu dark matter) and blue (μ=mu dark energy). This is also true for anti-red (cyan) and anti-blue (yellow).

The colors originate in quantum chromodynamics (QCD), the Standard Model for strong interactions since 1972. We added the letter designations and swapped anti- for subtractive CMYK colors.

$$\frac{\text{Poisson's gravity } \nabla^2\Phi=4\pi G\rho}{\text{Laplacian } \nabla^2} \Rightarrow \Phi_m=4\pi G\nu$$

1.9: Poisson's Gravity Function

Einstein's field equations convert to energy functions using the Gauss-Poisson equations method[27] to compute order=nu (ν). In the fundamental law of Thermodynamics, singularity occurs where S=0, $\delta\nu$=PdV/c², making disorder $\delta\mu$=TdS/c². The laws of Thermodynamics therefore define the phase moment (dU) in the Occam's Cycle.

Expanding void is the idealization of mu, where singularity is the idealization of nu. The color charge associations are associations with quantum numbers. The unit values are: dm= ($\delta\mu$=TdS/c²) − ($\delta\nu$=PdV/c²). The universe depends on inefficiency. This alone is too efficient.

---

[26] Nave, C.R. (2017). Eigenvalues and Eigenfunctions. Georgia State University. http://hyperphysics.phy-astr.gsu.edu/hbase/quantum/eigen.html
[27] Pe'er, A. (Feb. 17, 2014). Einstein's field equation. Cork Ireland: University College. http://www.physics.ucc.ie/apeer/PY4112/Einstein.pdf.
Varnes, E. (2004). More on Gravitational Fields and Potential. physics.arizona.edu/~varnes/Teaching/321Fall2004/Notes/Lecture20.pdf.

Inefficiency creeps in by means of the equivalent δQ±δW. Even if that could fail, the bubble mix guarantees anomaly.

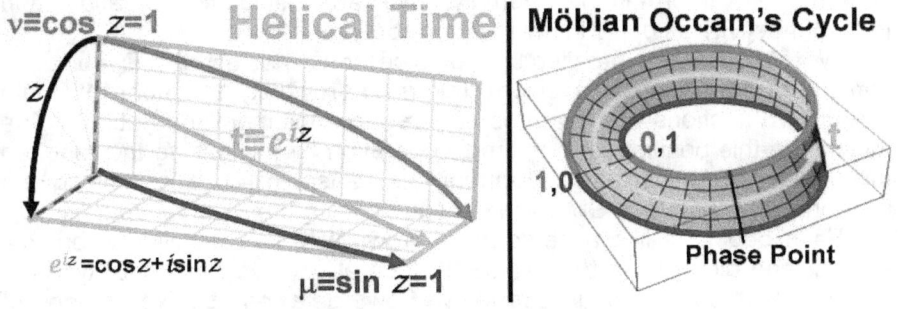

1.10: *Helical Time & Occam's Cycle*

Changes in pressure (P) and disorder (S) can describe either work (W) or energy going into a system (Q)—and they are split between the two sets of terms. Nothing specifically in TdS or PdV is exclusive to δQ or δW. We assume T goes into Q, but the energy entering the system could be intrinsic gravity and therefore not transferred like T.

There are several ways to write the function for deriving time from strong interactions like change in time occurs as increments of change occur between j and i: dt= δj + δi. Here we used Euler's helix formula.[28] This formula is commonly used to rotate the y axis with the imaginary i onto the x-plane resulting in a three dimensional space.[29]

Time is commonly assumed to be the change radian. The change radian on Occam's Cycle follows the phase point, so we used z to indicate that abstraction. The logical change function of imaginary i rotates phase relative to the time-emergent axis. It provides the sinusoidal twist/wave shape. Occam's Cycle time links by Quantum Shade to expansion of void, CMBR and Thermodynamic inefficiency (pg. 18).

Quantum Shade

Quantum Shade holistically connects all points requiring conservation and proportional opposite reaction in all dimensions. We focus here on time and space. Quantum Shade enables us to hold one variable class constant like space and look back into history, or time and see how value is simultaneously redistributed.

---

[28] Weisstein, E.W. (2018). Euler Formula. From MathWorld--A Wolfram Web Resource. http://mathworld.wolfram.com/EulerFormula.html.
[29] Joyce, D. (1999). Dave's Short Course on Complex Numbers: Multiplying Complex Numbers. Clark University. www2.clarku.edu/~djoyce/complex/mult.html.

It is as simple to conceive as a chalk board written upon in one moment, erased then written upon in the next moment. The content is impermanent information: anomaly. We are using conservation and interconnectivity as opportunity to do or observe.

We can use our architecture for clues on what a thing is doing. In computers this position is the microkernel, providing "the near-minimum amount of functions and features required to implement" operation.[30] The function in this position is to optimize operation by minimizing the required functions. This minimizing opportunity is causal of $\lambda\upsilon$=c explaining the horizon, Doppler effect, CMBR, and even defining the fabric of spacetime.

Each bubble is like a person in a crowd with its own history, origins, ending, and direction in the middle. It does all its own things but with the same laws. Observational astronomy shows galaxies at mixed stages of development at every redshift value. The ultra-deep field has a number of smudges deemed galaxies of various or unspecified developmental stages (e.g. CN-z11[31]) to a gamma burst.

Lyman-α galaxies occur in the full range of observations, even at redshifts practically next door (0.19 – 0.45).[32] These are new bubble systems creating massive quantities of new particles and hydrogen. Suggestive irregularity and morphology changes increase with distance at least up to a redshift of 4.[33]

Surveys of clusters and galactic interactions suggest even greater ambiguities.[34] Such ambiguities we will refer to as instantiations—existing or having qualities due to a particular context; a perturbation. To retain these, the universe cheats and optimizes its recycling. As an example, a primordial color charge must be confined (hidden in interaction).[35]

Real photons consist of a color charge interacting with its reflection, confining in a time-neutral entanglement. One change function is used twice for one identity with significantly less energy than two would require.

---

[30] Janssen, D. & Janssen, C. (2018). techopedia.com/definition/3388/microkernel.

[31] Newman, P. (Jul. 26, 2016). The Farthest Visible Reaches of Space. Goddard Space Flight Center. imagine.gsfc.nasa.gov/features/cosmic/farthest_info.html.

[32] Hayes, M. (2015). Lyman alpha Emitting Galaxies in the Nearby Universe. https://ned.ipac.caltech.edu/level5/Sept15/Hayes/paper.pdf.

[33] Conselice, C.J. (Jul. 21, 2004). The Galaxy Structure Redshift Relationship. ned.ipac.caltech.edu/level5/March04/Conselice/paper.pdf.

[34] Calvi, R. et al. (Nov. 15, 2011). The distribution of galaxy morphological types and the morphology–mass relation in different environments at low redshift. onlinelibrary.wiley.com/doi/10.1111/j.1745-3933.2011.01168.x/abstract.
Guan-wen, F. et al. (2016). Morphological Classification of High-redshift Massive Galaxies in the COSMOS/UltraVISTA Field. sciencedirect.com/science/article/pii/S0275106216300376.

[35] Greensite, J. (2011). An Introduction to the Confinement Problem. Springer.

Quantum Relativity

The energy doesn't sit still. It is swapped around among the parts and interactions in microstates thanks to cosmic void expansion.

On the grand scale the bubble parts also go through value exchange cycles. This is obvious on a local scale where we see galaxies interacting with the singularity they are nucleated around. Without these structures, the singularity enfolds out of existence. In theory, the singularity absorbs only part of its galaxy, emits that, and recycles.

On a grander scale, galaxies interact, collide, and singularities interact directly. This makes CMBR ambiguous since it has an unknown number of sources not always active. Interactions are temporal and do not change bubbles or proportions until annihilation. Since each bubble also belongs to countless other bubbles, terminating one simply leaves a density distortion. This is prominent where the black hole was (a kaleidoscope effect).

It is because all the bubbles are equivalent that we can say all singularities are to some degree connected to all void points in the universe. Their relevance is mostly local. One of the key things these bubbles do is create the generic linear fabric of spacetime. Again a myriad of bubbles together give a seemingly smooth spacetime, but with perspective glitches, like the Doppler effect. Add to this spacetime objects with mass and you get relativistic patterns of density distortions with holes where the singularities cut out their personal segments as below. [36]

1.11: Spacetime Fabric and Quantum Shade

The spacetime of a galaxy is the operational fabric sustaining a black hole—the focus of dilation. This is the space where the hole in the fabric creates a displacement that dilation contracts toward. The rest of the universe including the content of the galaxy toward the horizon is pulling

---

[36] Singh, K.P. (April, 2017). <u>Gravity as a wave in space-time fabric</u>. brilliant.org/discussions/thread/gravity-as-a-wave-in-space-time-fabric/

Quantum Relativity

the other direction—expanding relative to contraction focusing wavelength to the horizon.

The redistribution of enfolded value is excluded from the galaxy. The further removed from the galaxy, the more the effect of redistribution—an inverse relation to proximity. As a black hole enfolds, the rest of the universe unfolds proportionally such that total relative space is constant. Conservation simply shifts density around by transformations.

2. Void Basics

Void generically means: invalid (useless), completely empty space (unfilled or caused to be empty), empty (free from, lacking, vacant), or (in bridge and whist games) being dealt no cards of a particular suit.[1] We do not use the term vacuum as a degenerate body fills vacuum but has a permeable void.

Our definition for void is the **difference between a whole and its parts**. For the universe, cosmic void is **everything that isn't enfolded into its singularity parts**. Every other point is subject to the conditions of universal void. The degree of impact, however, is cause for this chapter.

Void, time, and light are infinitely divisible antithetical concepts. They are not fundamental. They are opposite to what creates them. Light is opposite to whole/form (e.g. matter). It has no form of its own making it environmentally dependent and observed virtually.

If parts make a whole, then void is the whole minus its parts. It is an availability of space, whereas vacuum is an absence of matter. Degenerate matter excludes other matter and fills vacuum, but also has void.

If the whole is defined by localized enfolding of spaces, then void is every point not enfolding. Time emerging from the bubble cycle resists change between order and disorder and applies exactly to those things. Time emerging from GFE satisfaction resists change in mass and by extension momentum.

Point into Dimension

Void is a space function and spaces do things. Here, void as a distribution is signified with the letter mu (μ) as a virtual blue/yellow color charge. This is not to be confused with the tensor symmetry symbols of the field equations or magnetic permeability. The spinning of the fabric of spacetime manufactures this dimension that transform into enfolding time into nu (ν) defining singularity.

This is suggestive of aether theory, but it is every and any point as befits field. To be proper matter, it needs to have a focused magnitude and be able to form a confining interaction. Otherwise it is a perturbation—a virtual particle we can observe as a CMB photon. It is a field.

Relativity is often described as a theory of gravity. More importantly, it is a theory of space doing things like gravity. Classical Use Mode thinking has a hard time accepting this, let alone the more complex QM. The result

---

[1] (2017). Void. <u>Oxford English Dictionary</u>. en.oxforddictionaries.com/definition/void.

is false attributions, like the idea that a mass shapes the spacetime around it, when really it partly shapes its own spacetime.

The GFE metric tensor ($g_{\mu\nu}$) is often incorrectly assumed to be intrinsic like $R_{\mu\nu}$ and $R/2$. It is a higher dimension that smoothes lesser conditions to work together[2] by asserting "how to compute the distance between any two points."[3] It is the cross-over point to smooth into extrinsic, as with Λ, the cosmological constant. Because Λ depends on variables outside the function, it is considered an arbitrary constant.

Our point here is to distinguish intrinsic and extrinsic spaces. We need to further distinguish what is void to what. The universe and cosmic void are both fundamental and higher dimensional. In set theory terms this is like saying it is the least possible subset AND the greatest possible superset at the same time.

This means cosmic void plays two vital roles in all material identities—all of their spaces. First, it gives a relatively undifferentiated distribution of value increasing a point of no dimension into dimension at a rate of c. Second, it provides what artists call a negative space—a difference between things, also called a Fermi surface.

Both of these applications of cosmic void are actively doing things. It is vital to observe these two roles when we get into things like strong interactions and microstates affecting intrinsic motion. The internal effects are relatively minor simply because they are modestly contained. The external effects are another story.

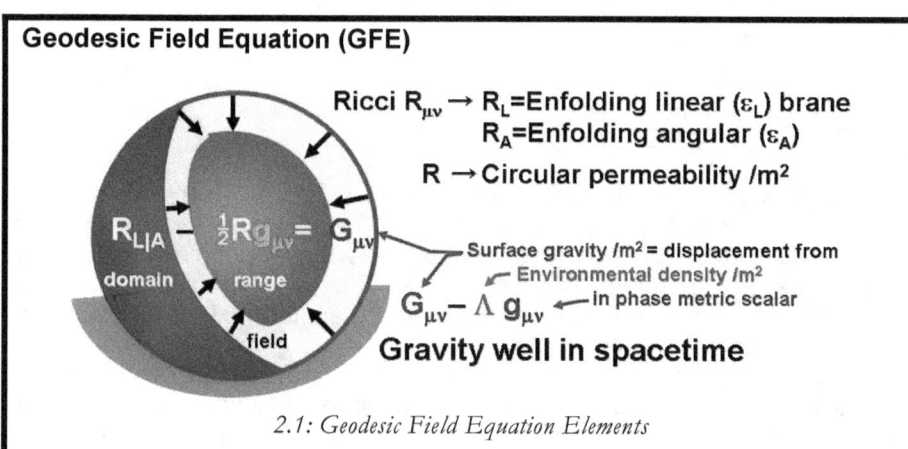

2.1: Geodesic Field Equation Elements

Relativistic manifolds are constructed from a quantum level. Their initial identities are lost in translation to surface branes (manifolds), but that

---

[2] Ranicki, A. (Dec. 2005). <u>High Dimensional Manifold Topology Then and Now</u>. University of Edinburgh. maths.ed.ac.uk/~aar/slides/orsay.pdf.
[3] Stover, C. and Weisstein, E.W. (2018). <u>Metric Tensor</u>. From MathWorld-- A Wolfram Web Resource. mathworld.wolfram.com/MetricTensor.html.

Quantum Relativity

is no consequence to surface gravity. It is significant in shaping the gravity well especially for singularities. The surface gravity is also the displacement from the environment, continuing the gravitational field via the local region of spacetime. This is why cosmological constant observations come back local to Earth.

The linear $\varepsilon_L = c^2/G$ and angular constants $\varepsilon_A = 1/4\pi\varepsilon_0$ are the quantum number values where singularity is triggered. These constants define relative proportions. They are permittivities, meaning they are what space allows. When one of these is achieved, the scalar curvature (R) becomes the available space for mass to permeate into.

Whatever is in R upon singularity formation is subject to satisfying the next quantum number. The permeation (filling) equivalent constants (G and $4\pi/\mu_0$) are the next quantum numbers. When these fill, the singularity discharges according to the chiral nature of its field.

R is divided by 2 due to being forced into phase shape, applying it to spherical volume by the metric tensor $g_{\mu\nu}$. Except $g_{\mu\nu}$, the tensors are all in $/m^2$ due to being put into common brane surface terms. This is vital for understanding how the EFE translate into quantum-functional terms making the EFE far more useful than originally planned.

Sea Level

This NASA WMAP[4] image shows CMBR distribution using the visible light spectrum from highest energy (violet) to lowest energy (red). You are the observer in the middle looking out to the inner surface of a spherical volume. Your position in the universe is always the center (another superposition quality).

2.2: NASA/WMAP CMBR Anomaly

CMBR is energy taken from strong interaction of matter annihilated and enfolded into singularities, and then unfolded generally to cosmic void.

---

[4] For current maps and details visit https://map.gsfc.nasa.gov/.

Quantum Relativity

This applies at the extremities of lowest fundamental and highest dimensional. Of course these interactions don't have the courtesy of being neatly distributed. The result is irregularity called anomaly.

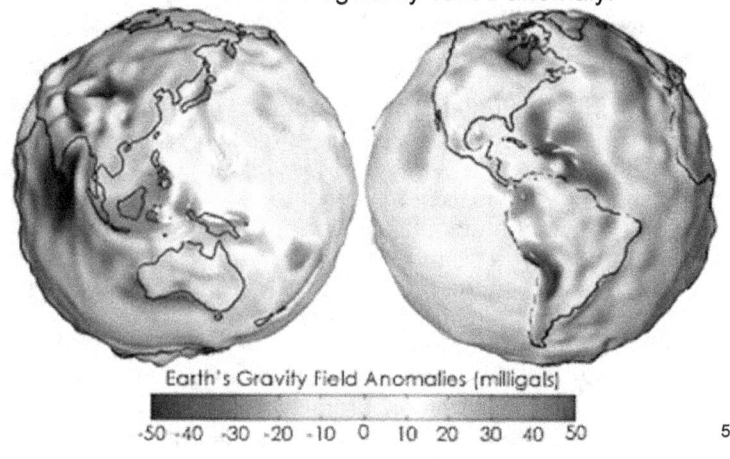

2.3: GRACE Gravitational Anomaly

Anomaly is another way to say disorder, and point to complexities. It occurs in every field except perfect order, as shown by the anomaly of Earth's gravity above. Sun spots are another example of anomaly. Like sun spots, the anomaly of CMBR varies over time due to complex causes.

With H_0=70.4, the average distance (radius) attributed to CMBR is about ~14 billion parsecs or 45.66 Gly (billion light years).[6] We are calling this "sea level" because it describes a midpoint about 97.58% of the way to the ideal horizon at ~46.85 Gly (see pg. 35). The difference percent-wise is about half the percentage of conventional mass: distributed resistance.

These diagrams are showing a surface. Take that surface in profile treating the WMAP like a topographical map. It looks like mountains and valleys. This is flux action—different force values applied at the same speed. The universe is pulling itself apart (blowing up) at rates up to the speed of light. Light defines the extremities, so c becomes the smoothed surface value and the medium of exchange on all levels of the universe.

In elementary chemistry, the concept of standard temperature and pressure (STP) is used to set a common frame of reference for observations. The standard is set by IUPAC,[7] where pressure is that at sea level. CMBR's "sea level" is the cosmic equivalent. It affects the distribution and patterns of galaxies resisted by their local interactions.

---

[5] Ward, A. (Mar. 30, 2005). Gravity Recovery and Climate Experiment (GRACE. earthobservatory.nasa.gov/Features/GRACE/printall.php.
[6] Gott III, J. et al. 2005. "A Map of the Universe." The Astrophysical Journal. 624 (2): 463–484. http://www.astro.princeton.edu/universe/ms.pdf.
[7] Nic, M. et al. (Feb. 24, 2014). IUPAC Gold Book: STP. goldbook.iupac.org /html/S/S06036.html.

Light & Photons

In the Standard Model, particles can be real or virtual (Gauge perturbations).[8] In this way, particles can be used as mediators of energy even though they technically don't exist.[9] The virtual particle system is unclear even among experts.[10] Virtual particles are **perturbations**, making them only real if they have enough value AND interact to confine.

The reasoning for this is simply that energy has to be put into the context of matter to observe it. Light is propagating energy commonly referred to as radiation like EMR. Radiation is awkward as a term too as it comes in soft and hard forms. Hard means the radiation can knock an electron out of its orbit.[11] Neither of these terms explicitly requires a real particle, though it is typically implied by "hard."

Light values but has no spacetime definition of its own. From valuing interacting spacetime constructs emerge other values. Light is given qualities from its source to its final destination. These qualities constitute information, such as absorption lines, which affect behavior. Absorption values contribute to the spectrum as attributed values.

They are attributed because the cause of absorption was either a potential taking energy out at a particular frequency value, or an intrinsic value obstructing the filling of value. These interference patterns are consistent due to general consistencies in microstates that affect light. All light in a field that is not otherwise modified, will have the same information. This is not to be confused with particle entanglement.

Light can behave both as a particle and as a wave propagation.[12] This doesn't help matters at all. Photons also have a bad habit of simply diffusing as light. We have to make the distinction though for the sake of creating matter and understanding observations. Light provides a scalar form of quantum force.

Photons are neatly packeted and focused. Light is not. If you left the aperture on the telescope open for hours, days or months to capture light, you observed a virtual photon. If it all arrived simultaneously in one focused point without any help, it was likely a real photon. Of course that means you can manufacture photons, but lasers were old news in the 80s.

---

[8] Nave, C.R. (2017). Feynmann Diagrams. Georgia State University. http://hyperphysics.phy-astr.gsu.edu/hbase/Particles/expar.html.
[9] Breinig, M. & Hitchcock, J. (2012). Physics 250: The Standard Model. http://electron6.phys.utk.edu/phys250/modules/module%206/standard_model.htm.
[10] Strassler, M. (2011). Virtual Particles: What are they? profmattstrassler.com/articles-and-posts/particle-physics-basics/virtual-particles-what-are-they/.
[11] Holman, G. & Benedict, S. (Aug. 1, 2007). What are Hard X-rays? https://hesperia.gsfc.nasa.gov/sftheory/xray.htm.
[12] Sevian, H. et al. (2000). Wave-particle duality. Boston University. http://physics.bu.edu/py106/notes/Duality.html.

Quantum Relativity

Expanding Void

The quantum universe is extremely deceptive. Among particles, it conceals its fundamental variables by requiring them to occur in proportions, be confined in complex identities, and finally degenerated to hide the true nature completely.

The fabric of spacetime is generated by such a myriad of bubbles simultaneously, that it provides a relatively smooth background to interact with. So if it is so neatly smoothed, why the massive Doppler effect at vast distances? Hubble provided the first plausible hypothesis.

In 1929, Hubble reported that distance accelerates objects away from us, concluding the universe is expanding.[13] Lemaître computed the universe to be expanding, which Hubble corrected using his information. At $H_0=500h$ Km/s Mpc, the universe is hardly expanding at all.[14] This constant is computed for general use by statistical inference of finite distance and local velocity information—$v=H_0D_L$ (see pg 34).

Since Hubble affirmed that galaxies are outside the Milky Way in 1936, a number of creative ways have been developed to check our methods of measure. Brightness is a good one assuming you know the original value. Cepheids were observed to be reasonably consistent. They became the go-to benchmark to compare the Dopler and distinguish motion from distance.

Supernovas and cepheid variables help distinguish individual velocities up to about 400 Mpc (1.3 Gly).[15] Let us assume these are perfectly reliable. These are unique, such that the "constant" is a measure of spacetime difference of a velocity-distance ratio. Combining as a general use number is a way to guesstimate.

By applying these creative methods to exhaustive observations, ideas began to form to help modify the Doppler z-shift and ascertain actual distance. One of these is to recognize that galaxies interact in a maze of patterns called filaments shaping also walls, and clusters.

This set of galactic interaction, is like a giant pot filled with spaghetti. Each of the pats is pulling its own direction. The whole tangled mess is being pulled vicariously apart by not just by local variables but impossibly distant variables.. The energy of the action translates as momentum (p) resisted by the mass of the strands of interaction (/m) provides relative velocity (v).

Of course this makes for a complex terrain of spacetime densities. More importantly, it creates a mass-like condition. This among other

---

[13] Hubble, E.J. (Jan. 17, 1929). A Relation Between Distance and Radial Velocity Among Extra-Galactic Nebulae. Proceedings of the Nat'l Academy of Sciences. apod.nasa.gov/diamond_jubilee/1996/hub_1929.html.
[14] Livio, M. & Riess, A.G. (Oct. 2013). Measuring the Hubble Constant. http://physicstoday.scitation.org/doi/full/10.1063/PT.3.2148.
[15] Freedman, W.L. et al. (2000). Final Results from the Hubble Space Telescope Key Project to Measure the Hubble Constant. arxiv.org/pdf/astro-ph/0012376.pdf.

interferences are given mathematical designations and estimated values. This was all initiated by Hubble to try and explain the expanding universe. Today, the expanding universe is the established view. But then we look at our bubbles and the very simple relationship between frequency and wavelength.

There is no doubt that all these interferences Hubble identified actually apply and affct the terrain light passes through. Of course each path will be unique. What isn't unique is what space is actually doing. While space is expanding at nearly all its points, it is proportionally contracting in local places to a degree that in total the universe is not expanding.

The expansion of space is a tad more complex than Schrödinger's cat. We can reasonably say the cat has been dead for decades and dismiss opening the box to see if it died by poisoning. We cannot so easily dismiss the concept and expanding space.

Yes, space is expanding indefinitely. And like the tread mill's conveyer belt, as one end expands the other converts and contracts. The forever expansion of space is stopped by its conversion into a function of time that dilates back to the enfolding origin. The real trick of the whole thing is the elusive concept of cycle. We arbitrarily apply it to time enfolding in frequency. It is a floating variable, meaning it equally applies to wavelength.

Classical $v=\lambda \upsilon$ evolved into its light application as $c=\lambda \upsilon$ with Maxwell. Unlike the classical environment-dependent velocity, the light version can create spacetime. Relativity then shows us how the function works. By leaving the concept of cycle a floating unit, we can see the length unfold away from the source and time enfold (dilate) toward the source. This inverse relationship shows a "sweet spot" roughly in the middle of the spectrum (below[16]).

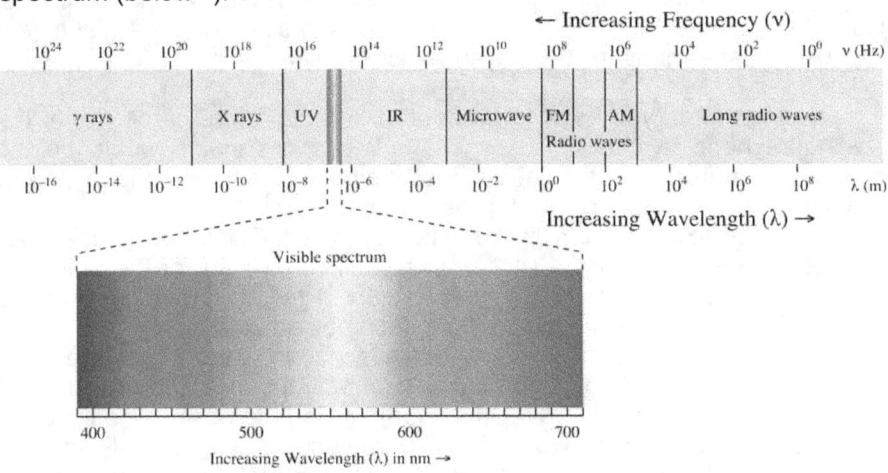

2.4: *Light Spectrum*

---

[16] DeMichele, T. (Sep. 29, 2016). Visible Light is Electromagnetic Radiation. factmyth.com/factoids/visible-light-is-electromagnetic-radiation/.

Quantum Relativity

The visible spectrum leans toward the dilation end of the spectrum. We should expect this since we evolved in this dilation region relative to the Milky Way singularity. It is the prevalent form of light here. As we move closer to the singularity, the effect will jump exponentially. Conversely as we go away, the cycle length increases while the increment decreases causing exponential wave length increase in the direction.

The reason the universe appears to expand is shown in this diagram just by recognizing the concept of dilation. That same concept applies in reverse to the length going the other direction. So yes and no. The universe is perpetually and proportionally expanding AND contracting in ways that end up defining a static universe. Our first warnings of this were the spectrum, conservation (requires finite and boundaries), and Einstein's static universe. Now we have the mechanisms to explain why.

Cosmic void expands at a constant c with an anomalous flux of CMB force subject to density. The latter is a result of the complex of bubble fields making prediction a QM problem to be solved. That expansion does many things. It does not mean the universe expands. Between the twist and enfolding, the expansion washes out. Typical quantum universe: depriving us of the easy solution and making things even more ambiguous.

Distance

Distance is the classical mensuration (measurement) problem of this chapter. The problem is seeing things from the perspective of a quantum universe. The universe gives us the perfect measure: light. Light cannot lie. All adjustments fit context.

Proper distance is the actual distance between two fixed points. Consider a map of Alaska. Assume the map is to scale. If it is a globe, you can use a tape measure to find an exact flying distance of 864 km (537 miles).

On a flat map, you have to account for the curvature of the Earth. If it were light bending over this space, the affected variable would be the omega curvature Ω_k. This flight distance we can classify as an ideal proper distance because it has no ambiguities.

The ideal full spectrum light horizon at 46.85 Gly is an ideal proper distance. Wherever you are is the superpositional center of the universe. The universe expands in every direction up to the speed of light. At the speed of light the wave function of light becomes a flat line.

This measure is ideal because it says nothing about the affects of terrain. Light can easily pass through a contracting spacetime that draws it into focus and causes it to travel much further than the observable limit.

Such a terrain issue would be like saying we will be walking from Anchorage to Nome instead of flying. All those mountains and valleys will definitely add to our distance. Let us say our traveler is carrying a device that measures every meter traveled.

Quantum Relativity

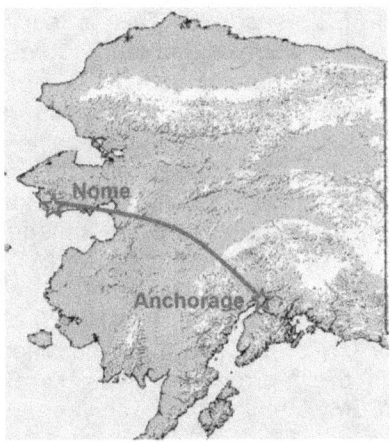

2.5: Anchorage to Nome

Astronomers are like an observer in Nome collecting the data from these devices coming in from every direction. These numbers are absolute observational values. In astronomy the key number is the redshift z value. The z value is used generically to compute velocity in cz=v.

Light Distance

Light distance is our traveler enduring the long walk through the variations in terrain. Terrain variations guarantee a longer path compared to flying distance. The reason is Relativity. Spacetime densities vary due to the material spaces passed through. These same spaces add curvature we would think of as terrain.

As observers stuck in one location, it is nearly impossible for us to reasonably identify distant terrains unless they are kind enough to throw up a light flag we can see. The inconvenient fact of observational astronomy is that we don't know. We only know the number we observe, and if we are lucky, it and the terrain are close enough that we can make out terrain and motion details.

Distance Functions

Uncomplicated light distance (D_L=cz; $z\lambda'=\lambda-\lambda'$) is a function of the ratio between observed (λ) and emitted (λ') values of wavelength. This is done by observing the shift of known absorption lines in the spectrum associated with specific elements. [17] The spectrum boundaries are numerically $h^{-2/3}$

---

[17] Wiggins, D. (Jun. 28, 2010). <u>Spectroscopy</u>. https://solarsystem.nasa.gov/deepimpact/science/spectroscopy.cfm.

=417 trillion years=1.3E22 seconds. This is subjected to the density conditions of the interacting bubbles and terrain to define the speed of light giving us the foundation of all our constants.

Cepheid variables have consistent brightness used as an alternative means of distance measure. [18] This brightness is used to help identify motion and compute the effects of density on expansion/dilation. Combining distance and velocity benchmarks up to 1.3 Gly provides our guesstimated terrain issues. These are the omega (Ω) modifiers: mass density (m), radiation (r), cosmological constant (Λ), and curvature (k) set at 0 assumes flat space.

The numbers provided below are with H_0=70.4[19] consistent with the established values for CN-z11.[20] The dramatic difference with H_0=100 or 500h Km/s Mpc is in the h-error that gives: D_L=6,467.4701Mly at 500 (lookback at 1,936.9726My).[21]

ΛCDM model Feb. 2010 Omega Data

Matter Density	Ω_m = 0.272
Radiation	Ω_r = 0.0000812
Lambda (cosmological constant)	Ω_Λ = 0.728
"Curvature"	$\Omega_k = 1 - \Omega_m - \Omega_\Lambda$ = 0
Hubble Constant	H_0 = 70.4 Km/s Mpc
Transverse Distance	$E(z)^2 = \Omega_r(1+z)^4 + \Omega_m(1+z)^3 + \Omega_k(1+z)^2 + \Omega_\Lambda$
Given a redshift of	z = 11.1
Proper Distance	$D_I = dz/dE(z)$ = 45.9337365 Gly
Lookback time	$t_I = \frac{1}{H_0} \int_0^z \frac{dz'}{(1+z')E(z')}$ = 13.7569076 Gy

2.6: Distance Elements & Calculations

Lookback time (t_L) is the standard way to measure distant objects. The idea is to take the transitive value based on when the photons were created.[22] The idea is sound, but it is also a terrain issue. The modifier was

---

[18] (Ret Feb. 23, 2018). Cepheid Variable Stars & Distance. Australia National Telescope Facility. atnf.csiro.au/outreach/education/senior/astrophysics/variable_cepheids.html.
[19] "70.4 ± 1.4 (km/sec)/Mpc" per: Wollack, E.J. (Mar 25, 2013). Tests of Big Bang: Expansion. https://map.gsfc.nasa.gov/universe/bb_tests_exp.html.
[20] Newman, P. (Jul. 26, 2016). The Farthest Visible Reaches of Space. Goddard Space Flight Center. imagine.gsfc.nasa.gov/features/cosmic/farthest_info.html.
[21] Jordaan, B.A. (2009). Relativity 4 Engineers. Ebook. Calculator: www.einsteins-theory-of-relativity-4engineers.com/cosmocalc.htm.
[22] Hogg, D.W. (2000). Distance measures in cosmology. Princeton, NJ: Institute for Advanced Study. https://arxiv.org/pdf/astro-ph/9905116v4.pdf.

already applied for the transverse distance it derives from. This is an over-application of modifiers

Secondly, photon creation is assumed within the reference frame of ~13,757,283,900 year ago. When you look for a number, you set yourself up to find it, correctly or not. The **photon's redshift tells you exactly how old it is**.

Third, another test.... Using $H_0=70.4$ with CN-z11 at $z=11.1$ (45,933.7365Gly),[23] IOK-1 at $z=6.96$ (45,564.4571Gly)[24] and their difference 369.2794Mly (only .5 My lookback difference!). This is next door. Let us analyze the logic of z-functions that incorrectly concludes $z>1$ means $v>c$[25] and explain apparent dilation at extremes.

$$z \lambda' = \lambda - \lambda' = \delta\lambda$$

First, z is a unitless measure of drift—the increment of change in frequency ($\delta\upsilon = \upsilon' - \upsilon$). Drift is exposed as the defining feature by unit discrepancy in $\lambda=(c\pm v)/\delta\upsilon$.

Drift translates frequency change into a duration ($1/\delta\upsilon$) and length ($c\delta\upsilon=\delta\lambda$). This duration and length are proper measures under condition of no resistance to void expansion. It solves the unit discrepancy, cancels in translation, and gives co-velocity relative to c: $v=\delta\lambda\delta\upsilon$. This expansion of the universe is limited to the total drift potential from one end of the light spectrum to the next so all light flatlines at

$$z = \delta\lambda/\lambda' \rightarrow cs/m$$

Resistance to expansion prevents direct conversion into distance. Decrease in resistance to expansion increases the drift rate. At $z=cs/m$, $H_0=70.4$ we calculate $D_L=46.85$Gly as the light horizon and $t_L=13.757$Gy (BBT). Publishing z with inadequate explanation is confusing. The cause is not dilation, Minkowski, or comovement.

The cause of exponential z increase is twofold. First, terrain influences affecting the propagation relative to the observer. Second, the light from any source is subject to the expansion of void relative to the light horizon of that source opposed by the enfolding of frequency.

The further from the source, the less temporal dilation affects wavelength causing exponential growth in z value. Reducing resistance shortens the distance to observation (depth). The surface is an abstraction, so the size and distribution of points on that surface are simple distances. And of course, with superposition, each of the points you observe at that horizon has a similar view to yours.

---

[23] Oesch, P.A. et al. (March 3, 2016). <u>A Remarkably Luminous Galaxy at z=11.1 Measured with Hubble Space Telescope Grism Spectroscopy</u>. spacetelescope.org/static/archives/releases/science_papers/heic1604a.pdf

[24] Hogan, J. (2006). <u>Journey to the birth of the Universe</u>. Nature. https://www.nature.com/articles/443128a.

[25] Wright, E.L. (Feb. 23, 2002). <u>Doppler Shift</u>. astro.ucla.edu/~wright/doppler.htm.

Comovement

The effect of the omega modifiers across the terrain is called the transverse comoving distance. Transverse is a fancy word for across. Comoving means as it suggests: two objects with common velocity on parallel and otherwise equal paths. Comovement as a variable does not reduce the distance or time light traveled. It can affect the current distances between objects.

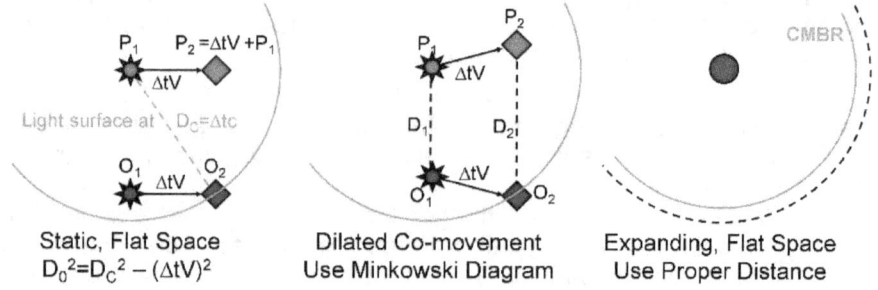

Static, Flat Space
$D_0^2 = D_C^2 - (\Delta tV)^2$

Dilated Co-movement
Use Minkowski Diagram

Expanding, Flat Space
Use Proper Distance

2.7: Co-movement v. Expanding Void

As light traverses greater distances, the expansion of void stretches the wavelength and diminishes the flow of energy (flux=S) and luminosity ($L = 4\pi D_L^2 S$).[26] A standard luminous reference like a cepheid or supernova provides a convenient frame of reference to compare the Doppler and distinguish expanding void from local velocity and direction. Andromeda is so local, blueshift alone indicates it is on a collision course.[27]

The problem of what to use when boils down to proximal relationships. Imagine a round table surface. You have two objects moving at the same velocity on parallel paths (first part of the diagram above). The observer (O) sees the history of light from the other moving point (P). Granted it is a pretty recent history. Next, numbers in fractions of the speed of light are definitely subject to dilation. Special Relativity applies—use Minkowski.

Important to note that time (t), distance (x), and velocity (v) are reduced to unitless fractions of the speed of light. This makes for easy conversion into circular functions.[28] Original values are indicated by primes (x', t') as opposed to current values. As $v \to c$, $t \to 0$, but at $v \geq c$, t is empty set and distance enters space-like realm.[29]

---

[26] Pogge, R. (2006). The Cosmic Distance Problem. Ohio State University. http://www.astronomy.ohio-state.edu/~pogge/Ast162/Unit4/cosdist.html.

[27] Phillips, T. (May 31, 2012). Astronomers Predict Titanic Collision: Milky Way vs. Andromeda. Science@NASA. https://science.nasa.gov/science-news/science-at-nasa/2012/ 31may_andromeda.

[28] Minkowski, H. (2012). Space and Time: Minkowski's Papers on Relativity. Minkowski Institute. http://rgs.vniims.ru/books/spacetime.pdf.

[29] Asmodelle, E. (Aug. 8, 2016). Pseudo-Orthogonal 4D Representation of Minkowski Spacetime. https://www.linkedin.com/pulse/pseudo-orthogonal-

Quantum Relativity

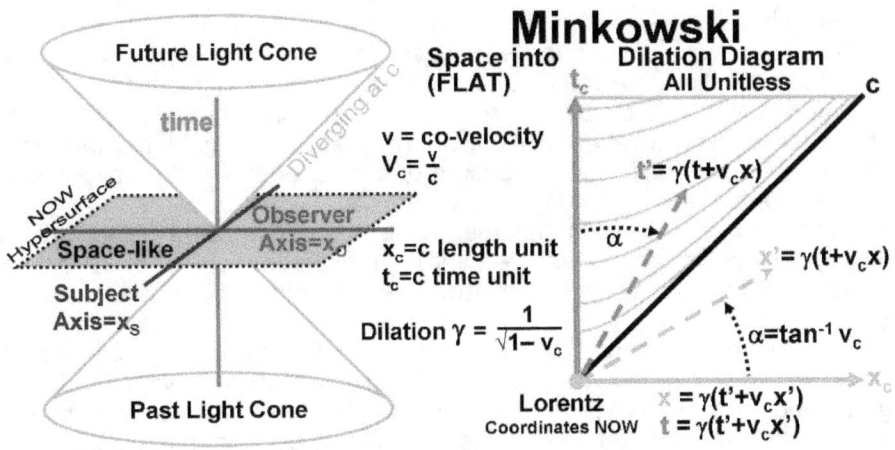

2.8: *Minkowski Space & Dilation Diagram*

Into Shade

Void is the dominant feature of the left-handed universe[30] of disorder. It is commonly called **dark energy**. We give it the material variable designation of μ (mu). This is misleading as points and expanding void are technically virtual mu.

Void makes up about 73% of value in the universe, which includes the 5 percentage points attributed to conventional mass. If we lump all the dark matter (ν=order) into a unit value (1), the dark energy with conventional matter would be e for an order to disorder (ν:μ) ratio of 1:e (27:73).[31]

Void is the transport layer of our architecture, meaning it conveys values between USE Mode (the range of actual matter) with PHASE Mode (the action of order sustaining disorder). Order needs the USE Mode and workable matter to act, and that action feeds into disorder.

Put another way: Space does things. Void specifically conveys and distributes value. This distribution makes everything in the universe work. Without its distributing and expanding, nothing happens. No time. No change. Nothing.

The speed of light as a constant derives from void expansion and the cosmic clock. As mentioned before, constants aren't just numbers, they do

---

4d-representation-minkowski-estelle-asmodelle/.
Evensen, K. (2009). An interactive Minkowski diagram. trell.org/div/minkowski.html.
[30] Egede, U. (May.11, 2006). Rare Decays at LHCb. London: Imperial College. https://www2.ph.ed.ac.uk/event-resources/ppe/2006/egede.pdf.
[31] (Jan 18, 2018). Dark Matter. CERN. home.cern/about/physics/dark-matter.

Quantum Relativity

things. These are not to be confused as reflecting CMBR flux (e.g. luminosity). CMBR value can never describe a larger universe. It can only describe a smaller universe (sea level) due to irregular distribution.

CMBR flux affects the force of expansion and the rates matter is created and evolved even in isolation. This acts as a hidden variable in decay series we could never measure or predict except statistically. It can also distribute unevenly due to local potentials.

Cosmic void expansion, the speed of light, and the cosmic clock are all set by the cycle and conserved through interconnectivity of Quantum Shade enabling the rest of Phase Mode. As void expands, light permeates, time increments.

The universe depends on anomaly to function. Bubbles cannot occur in isolation. They overlap, interact, and by degeneration in vast numbers, guarantee smoothing of anomaly. Due to their interconnectivity, the entirety of the universe ultimately evaluates as a single cycle of microstates in a sequence. And because the universe behaves optimally, we can make a reasonable and shocking hypothesis of this cycle.

The universal cycle arguably means the entire universe already happened. And it only had to happen once, just like any other identity. Once that cycle is defined, it just is. Meanwhile, we weary travelers just happen to be stuck trying to work things out navigating the pages of history the slow way around. No guarantee that this pattern in any way is relevant to us or us to it. We are insignificant even on a local level.

The inconvenient truth is that Bell's theorem applies. We can't observe the hidden variables. There are too many observable variables to simultaneously observe. The universe is one giant QM problem, and there is no way we can possibly acquire or identify all the variables, values, or computational power to calculate them. We must treat it as a QM problem or forever be chasing houses on the winds of tornadoes in Wonderland. [32]

---

[32] Bell, J.S. (Nov. 4, 1964). <u>On the Einstein Podolsky Rosen Paradox</u>. Madison, WI: University of Wisconsin. cds.cern.ch/record/111654/files/vol1p195-200_001.pdf.

3.   Creating Singularities

The concept of singularity originates in mathematics where it describes the breakdown of a mathematical object or a misbehaving exceptional condition—wherever an equation "blows up" and becomes degenerate.[1] We tend to think of singularities in relativistic black hole terms.

Einstein's Special Relativity emerged in two of his four 1905 science journal articles.[2] Over the next ten years, his ideas soaked into the scientific community while he worked out the General connection. From General Relativity came cosmology in 1917.[3] This was a static universe that is spatially finite, with a uniform distribution of matter, introducing the concept of gravitational singularity.[4]

Einstein received the Nobel Prize for his contribution to the photoelectric effect, which kicked off the pursuit of quantum theory.[5] Despite his plethora of contributions to quantum theory, he dismissed QM with "God does not play dice." Einstein's gravitational singularity abandons the logic solving the problem—a misunderstanding. Quantum asserts the mathematical logic and works with the observations (e.g. Planck bounce). All singularities here are quantum/mathematical.

Mathematical Root

The 18th Century successors of Leibnitz (1646–1717) and Newton (1642–1727) waged a mathematical revolution. They included Laplace, Legendre, Lagrange, Euler, the Bernoulli's, L'Hopital, Taylor, Fourier, Poisson, the next generation of Gauss, Fourier, Herschel, Ampère, et al.[6]

---

[1] Weisstein, E.W. (2018). Singularity. From MathWorld--A Wolfram Web Resource. http://mathworld.wolfram.com/Singularity.html.
[2] Panek, R. (Jun. 2005). The Year of Albert Einstein. Smithsonian. smithsonianmag.com/science-nature/the-year-of-albert-einstein-75841381. Einstein, A. & Minkowski, H. (1952). The Principle of Relativity. Dover.
[3] Einstein, A. (1917). Cosmological Considerations in the General Theory of Relativity. Trans. Perret, W. & Jeffrey, G.B. einsteinpapers.press.princeton.edu/vol6-trans/433.
[4] O'Raifeartaigh, C. et al. (Jan. 25, 2017). Einstein's 1917 Static Model of the Universe. https://arxiv.org/ftp/arxiv/papers/1701/1701.07261.pdf.
[5] Darling, D. (2016). Einstein and the Photoelectric Effect. daviddarling.info/ encyclopedia/E/Einstein_and_photoelectric_effect.html.
[6] Wilkins, D.R. Mathematicians of the Seventeenth and Eighteenth Centuries. Dublin, IR: Trinity College. maths.tcd.ie/pub/HistMath/People/RBallHist.html.

Practically anyone who was anyone in the world of mathematics you see prominently in advanced mathematics textbooks today lived in the 18th Century. This includes Partial Differential Equations, the original concepts that would evolve into manifolds, and the first of what we can consider modern field equations like Poisson and Gauss's gravity.

To prove his profound ideas, Einstein needed a way to link his thinking with established physics. Gauss and Poisson paved the way to connect Special Relativity thinking to Newton's Inverse Square Law.[7] Einstein's Relativity simply evolved a system of thought with already deep roots.[8]

No one was ready to properly analyze the singularity Einstein saw in his field equations. The first clue predated Einstein. Planck's "Black Body Theory" of 1900 is typically where a history of quantum theory begins.[9] The perspective didn't really take off until Einstein's photoelectric effect, likely due to his popularity in an emerging video news media.

It wasn't until Schrödinger's cat (1935[10]) that the concept of superposition was introduced to begin explaining what Einstein was looking at. By this time, Relativity and Quantum Theory had established separate camps, leaving singularity in the wrong camp. Singularities are definitely a quantum phenomenon, not relativistic.

Quantum Approach

Einstein's geodesic field equation (GFE) emerges into an evolved form of the Poisson-Gauss field equations for gravity. This enabled him to transition from GFE and spacetime doing things (field theory) into Newton's Inverse Square Law.[11] It also revealed boundary conditions where classical gravity breaks down as its variables quantize.

Einstein evolved the Poisson-Gauss function by using the Schwarzschild radius ($r_S = 2MG/c^2$),[12] with M being in the $T_{\mu\nu}$ manifold. This specific function ($8\pi G T_{\mu\nu}/c^4$) is space curving in to shape the field of gravity. The Schwarzschild radius is a form of permittivity—what a spacetime will allow=ε, see pg. 56)[13]. It is the ground state enfolding value

---

[7] Brown, K. (2017). Poisson's Equation and the Universe. mathpages .com/home/ kmath711/kmath711.htm.
[8] Petrov, V.A. 100 Years of Relativity Crucial Points. Protvino Russia: Div. of Theoretical Physics, Institute for Higher Energy. web.ihep.su/library/ pubs/ tconf05/ps/c5-2.pdf.
[9] Cresser, J. (2009). The Early History of Quantum Mechanics. Macquarie University.
physics.mq.edu.au/~jcresser/Phys201/LectureNotes/EarlyHistory.pdf.
[10] Schrödinger, E. (Nov. 1935). The Present Situation in Quantum Mechanics. Naturwissenschaften. https://doi.org/10.1007/BF01491891.
[11] Carroll, S.M. (Dec. 1997). Lecture Notes on General Relativity. UC Santa Barbara. https://arxiv.org/pdf/gr-qc/9712019.pdf.
[12] Schwarzschild, K. (1916). About the Gravitational Field of a Mass Point According to Einstein's Theory. German Academy of Sciences in Berlin.

of a singularity. The order/permittivity space is what the angular pressure of $T_{\mu\nu}$ curves into.

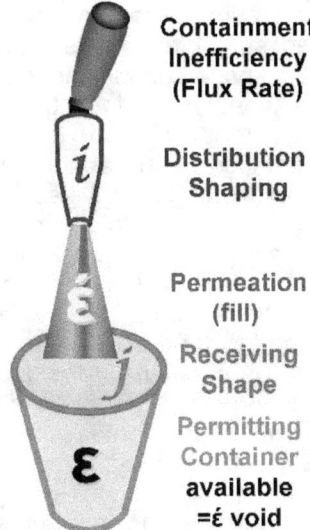

3.1: *Tap Shape Permeates fill into Cup's Permitted Shape*

The electromagnetism concepts permittivity and permeability apply to other contexts. Newton's constant is a linear form of permeability—the effect of energy filling a spacetime. The permittivity-permeability product $\acute{\epsilon}\epsilon=c^2$—defines a fully occupied spacetime. Since each of these is a quantum number, filling both triggers Pauli exclusion.

For a singularity, permittivity is set by a radiant value (EMR) and defines the enfolding space. The second field is the space made available for value to fill. This is an attributed value (EMA) space defined by permeability. When that permeability value is filled, the quantum number violation triggers exclusion and ejection (quasar).

$T_{\mu\nu}$ is a linear pressure value (kg/m s²)[14] curved into a spherical angle (4π) in a spacetime (/c²). Angular permeability derives from magnetic permittivity, so: $4\pi T_{\mu\nu} \rightarrow \acute{\epsilon}_A = 4\pi\epsilon_0 c^2$. Dividing by c^2 reduces this to mass per volume (kg/m³). It is disorder flowing into an ordered surface.

Singularity derives from perfectly ordering (S=0) a density degenerate (see pg. 172). Value pushed into a singularity's permeability space does not change the density. It does increase the radius of the singularity displacement and related gravitational the intensity feeding into it. Expelling the value bounces the radius to ground state value.

---

[13] Bevelacqua, P. (2012). Maxwell's Equations: Permittivity. maxwells-equations.com/materials/permittivity.php.
[14] Myers, A.L. (2016). Natural Units of General Relativity. University of Pennsylvania. https://www.seas.upenn.edu/~amyers/NaturalUnits.pdf.

Quantum Relativity

Einstein tripped on the boundary conditions and realized the EFE could be read classically or as a singularity. Here he stumbled despite this specific function showing the conversion into a brane. A **brane** is a surface with no depth or specified shape/form. It is created as a mathematical abstraction to put various spatial definitions into a common frame of reference. By doing this conversion into the abstract, Einstein was divorcing conventional spatial reasoning.

Degeneracy in mathematics is an abnormal condition. We expect gravity to behave relative to the origin of a geometric solid like a sphere. We don't expect it to behave relative to the origin of a surface lacking geometry. When you take the geometry away, all the points are the origin with identical pressure value. Having identical points means there is no variability of information in the system. A lack of information variability is perfect Thermodynamic order: S=0 and T=0.

The units for $T_{\mu\nu}$ are for pressure. At singularity, this pressure is confined to the brane. It has no range. The intrinsic gravity of a singularity is confined. Its displacement effect comes from putting the brane into phase perspective. More importantly, as pressure, $T_{\mu\nu}$ is the enfolding action. For action, you need time.

Time technically only reaches the perfect suspension traditionally attributed to singularity at the geometric origin. It is only suspended so long as there is disorder for the singularity to interact with. The geometric origin is the only place definitively intersecting the enfolding brane.

Time dilates toward the singularity, but the disorder around and permeated into the singularity reduces the rate of dilation. Accumulated mass is confined by the brane, increasing the temporal rate and function of equilibrium. Time clearly plays very important and active roles in a singularity. Disorder resists dilation and prevents enfolding action from outright negating itself.

Quantum gravity is this primordial gravitational surface (brane). To find common variables for his geodesic manifolds, Einstein created branes. Branes are either in phase (shaped onto a volume) or out of phase dependent on context.

Light cannot interact with a singularity because the brane from that context is out of phase. It offers nothing for light to interact with. Light propagation, however, is temporal. From the temporal perspective, the brane is put into phase. This displaces a volume of space the propagation bends around.

This same displacement is the "hole" in the fabric of cosmic spacetime. The displacement relative to the pressure of the environment defines microgravity with $G_{\mu\nu} - \Lambda g_{\mu\nu}$ (for discussion of all gravity phenomena, see Relative Field Theory, pg 135 et seq.). Microgravity is the environment creating the long-range gravitational effect. This is one effect used to compute mass. Singularities do not have intrinsic mass, only contextual mass. A singularity is an ideal order perturbation, a virtual identity, faithfully obeying the laws of Thermodynamics.

Thermodynamics

Introduced in the 19th Century, Thermodynamics is indisputable practical engineering and an advanced topic. It is the key change element to all material changes, including the universe in total. Let us examine the three core laws and enhance them with Perkins' conversion and Fleming's sequence. We ignore the "zeroth law" (algebraic commutation: if a=b and b=c then a=c) and competing 4th law suggestions.

1. **Conservation**—Part but not all energy (efficiency) put in a system (δQ) converts to work (δW), the totality increasing the energy of the whole ($dU = \delta Q \pm \delta W$).[15]

    δQ enters a system opportunistically as: intrinsic to matter, radiant light, transferred or shared by interaction. Work on the system is $dU = \delta Q + \delta W$ equivalent to relativistic momentum. Work done by the system is $dU = \delta Q - \delta W$.

    Never assume $TdS - PdV$ translates directly into $\delta Q \pm \delta W$ in a complex system. Work is a subjective term, especially if your work is producing heat or diversity, or the energy acting on the system is intrinsic PdV (e.g. gravity). Contextual opportunism emerges in the 4th law, dividing things whichever way is most expedient. The emergence of a singularity in the 3rd Law is another loophole.

2. **Multiplicity**—Disorder (S) increases in an isolated system toward "thermal" (T) equilibrium.[16]

    Read "thermal" as transferable, linking change in T to volume (dV) and dS to pressure (P) in an isolated system. True also in a symbiotic system, except dS and V can increase for one part of the system as they decrease for another. Multiplicity=diversity of change functions (classic entropy, QR, QCD) and information sequence.

    Equilibrium here clearly means energy is being absorbed or emitted. It can also mean an imbalance of normalized (smoothed) distribution. Such an imbalance can certainly be causal of emission or absorption.

3. **Reduction**—A "perfect crystal" at absolute 0=T has no disorder (S=0). More simply: at $dU = PdV$, $T = dS = 0$.[17]

    $T = dS = 0$ is thermodynamic singularity. Matter is termed degenerate when disorder approaches zero ($S \to 0$). An isolate creates a singularity by regionally focusing multiplicity (5th Law). Singularity then consumes the remaining isolate, strongly interacts with some of that consumption and emits the annihilation triggering the LAE process.

---

[15] Kreidenweis, S. (2011). The First Law of Thermodynamics: Conservation of energy. chem.atmos.colostate.edu/AT620/Sonia_uploads/ATS620_F11_Lecture2/Lecture2_AT620_082411.pdf

[16] Lintner, B.R. (2015). Thermodynamics of the Atmosphere: Lecture 12. Rutgerss University. envsci.rutgers.edu/~lintner/thermo/Lecture12.pdf.

[17] Kyle, B.G. (1994). The Third Law of Thermodynamics. ufdcimages .uflib.ufl.edu/AA/00/00/03/83/00123/AA00000383_00123_00176.pdf.

Quantum Relativity

4. **Conversion**—Helical changes[18] (i) to radiant ($-\hbar E_1 \equiv pc$) energy apply degrees of work conveniently to AND by the intrinsic definition ($-i\hbar E_1 \rightarrow j E_0 \equiv mc^2$). See image below.

$$\{ [dU = \hat{\imath} E_2] = [\delta Q = -\hbar E_1] \pm [\delta W = \hat{\jmath} E_0] \}^2 \equiv \{ (pc)^2 + (mc^2)^2 \}$$

Combines Euler's $z = x + \hat{\imath}y$,[19] Relativistic Momentum,[20] Parallelogram Law,[21] Perkins' ice maker,[22] Conservation,[23] Fleming's rules,[24] logical operators (later), and the 1st Law. The complex components make work on OR by the system equivalent in Relativistic Momentum as ($\hbar^2 \equiv \hat{\imath}^2 \equiv \hat{\jmath}^2$) = −1. Relativistic equivalence makes OR into a simultaneous AND.

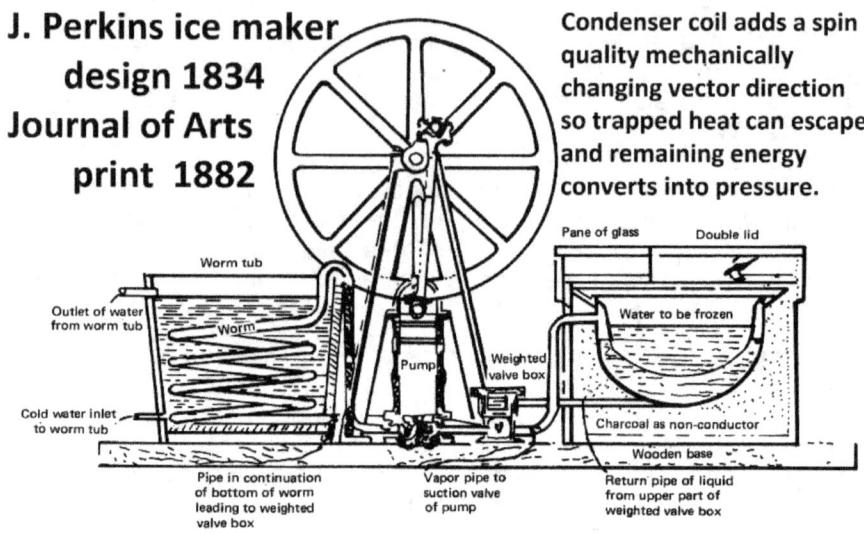

J. Perkins ice maker design 1834 Journal of Arts print 1882

Condenser coil adds a spin quality mechanically changing vector direction so trapped heat can escape and remaining energy converts into pressure.

3.2: Perkins & Rotating Heat into Cold

Work is distributed conveniently (e.g. path of least resistance) in degrees to the intrinsic definition (absorbed) AND by the intrinsic definition

---

[18] Not to be confused with a heat sink using 2nd Law thermal conductivity to redistribute heat away from a body.
[19] Song, H.A. (2008). Euler's Equation. songho.ca/math/euler/euler.html.
[20] Nave, C.R. (2017). Relativistic Momentum. Georgia State University. http://hyperphysics.phy-astr.gsu.edu/hbase/Relativ/relmom.html#c1.
[21] Jeffreys, H. and Jeffreys, B.S. (1988). Methods of Mathematical Physics. 3 ed. Cambridge, England: Cambridge University Press.
[22] ASHRAEnews. twitter.com/ashraenews/status/895618941887827969/photo/1
[23] Noether, E. (1918). Invariant Variation Problems. Nachr. D. König. Gesellsch. D. Wiss. Zu Göttingen, Math-phys. Klasse. pp 235–257.
[24] Daware, K. (2014). Fleming's Left Hand Rule And Right Hand Rule. electricaleasy.com/2014/03/flemings-left-and-right-hand-rule.html.

(emitted). The universe given two solutions by its own rules finds a way to apply both, visible in spectrum analysis.

Treated as a generic mechanical operator, the imaginary number (i) rotates axes to make linear and sinusoidal systems compatible in Complex Variables (the mechanical roots of Differential Equations). It works mathematically AND logically the same as the imaginary operators used here to define color change functions. When the operator disappears, the logical function remains to haunt buried in the definition.

5. **Hierarchy**—Field axes of the same magnitude are at right angles to each other,[25] ordered by scale from lowest to highest: gravity (g), thermal/heat (T), centrifugal (C'), centripetal (C), electromagnetic (e).

Order is set by following the value sequence in Fleming's rules. It demands an evolving system of fields/matter. This law is also a function of Relativity forming degeneracy toward singularity in a stellar body[26] and derives from the 2nd Law.

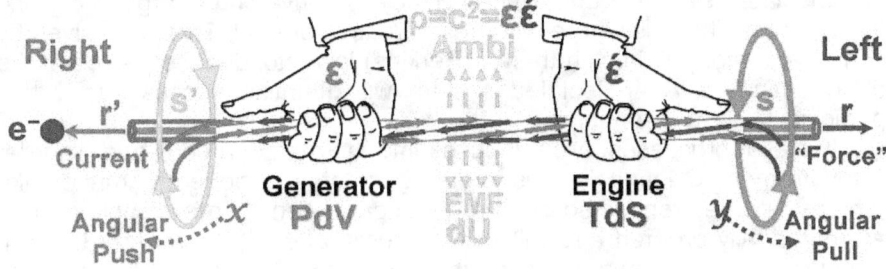

3.3: Fleming's Rules & Force Variables

Fleming's right and left hand rules show energy direction, generation, application in a working engine,[27] and its ambient loss. The diagram breaks down quantum forces for a more detailed analysis of chromodynamics. Ordered forces r' and y correspond with gravity and centripetal; r and x with heat and centrifugal disorder; spinors s' and s are both.

The generator converts a disordered value for transport by an ordered conveyance. Since this is an excess of the conveyor's identity, it is lost to other potential conveyors or as light (EM) emission. Loss happens anywhere in the system. Some disorder gets converted into subjectively defined work. The work can be order, disorder, or both.

Fleming's rules gives us a natural sequence putting order before disorder—gravity before heat. If we are thinking about a solid-state thing, the inefficiency will apply generally to the whole, putting the EM both last. The main vectors in their solid state sustainable order are gravity (order), heat (disorder), centrifugal (disorder), centripetal (order), and EM (both).

---

[25] Parallelogram Law.
[26] Mastin, L. (2009). Creation of Black Holes. physicsoftheuniverse.com/topics_blackholes_blackholes.html.
[27] Daware, K. (2014). Fleming's Left Hand Rule And Right Hand Rule. electricaleasy.com/2014/03/flemings-left-and-right-hand-rule.html.

Quantum Relativity

This is important for creating a quantum singularity thermodynamically and explaining its two fields. The universe always takes the path of least resistance—and to degrees all the others too.[28] The path of least resistance is to follow this sequence (e.g. change information).

Creating Singularities

Einstein tried to avoid wrapping his head around singularities too much. We can't blame him. The functions can be interpreted from the variable perspectives of linear (surface) or angular (volume), and in how they are created. Singularities are created at any scale by quantum fluctuation contributing to the perfect ordering of information in a degenerative thermodynamic process.

The second law of Thermodynamics states that disorder (S) increases in an isolated system[29] due to inefficiency of converting energy into work. What it doesn't say is where disorder increases or how. This is resolved by converting information with the 4th (Perkins) law into ideal sequence of the 5th (Fleming) law. An isolated system will quantize excess into virtual particles, like Weyl fermions.[30]

In stellar processes, order focuses into the core of a body and disorder away from it. Generically this ordering creates degenerate matter like protons and neutrons. Hadrons as degenerate density are complex. They are not ideally ordered and will actively resist attempts to reorder them. If the star is big and energetic enough, its core will form a field potential for degeneracy. The trick is to keep it active while it is finding a sense of order, without forcing it into quantization.

It is easy to look at permittivity and permeability and think all we need is to satisfy a quantum number. We also need the sequence. Without the sequence, we have a confusing state of degeneracy where the space is filled, yet there is a quantum dimension of it available without specific locations. This abstract availability pattern Wheeler called quantum foam (see pg. 172).

When we come short of the ideal sequence, the result is a neutron star/pulsar. The way to tell the difference is the context. A neutron star can survive without context to prevent annihilation by enfolding itself. It can pull in and eject as a gamma ray burst the entire body creating it in a matter of seconds. It can then continue without recovering any of that value or having other disorder around to sustain it. A singularity requires a context that will help it feed back into itself.

---

[28] Holt, M. (Jul. 1, 2001). The Path of Least Resistance. ecmweb.com/content/path-least-resistance

[29] (Nov. 5, 2016). 2nd Law of Thermodynamics. University of CA, Davis. chem.libretexts.org/Core/Physical_and_Theoretical_Chemistry/Thermodynamics/ Laws_of_Thermodynamics/Second_Law_of_Thermodynamics

[30] Xu, S. et al. (Aug. 7, 2015). Discovery of a Weyl fermion semimetal and topological Fermi arcs. science.sciencemag.org/content/349/6248/613.

Quantum Relativity

Let us assume the star is massive enough and has the right energy and density needed to sustain the field conditions through to ideal sequence. There are two chiral types of black hole. Chiral means imperfect mirroring. This imperfect mirroring is the quantum equivalent of antimatter. If these two try to occupy the same space, they will not join. They will systematically annihilate each other in very slow motion.

$T_{\mu\nu}$	Angular stress-energy	kg/m s²	Pressure parallel to surface=$f(\varepsilon_A)$
$C_{\nu\mu}$	Linear stress-energy	m³/kg s²	Volume mass density =$f(G)$

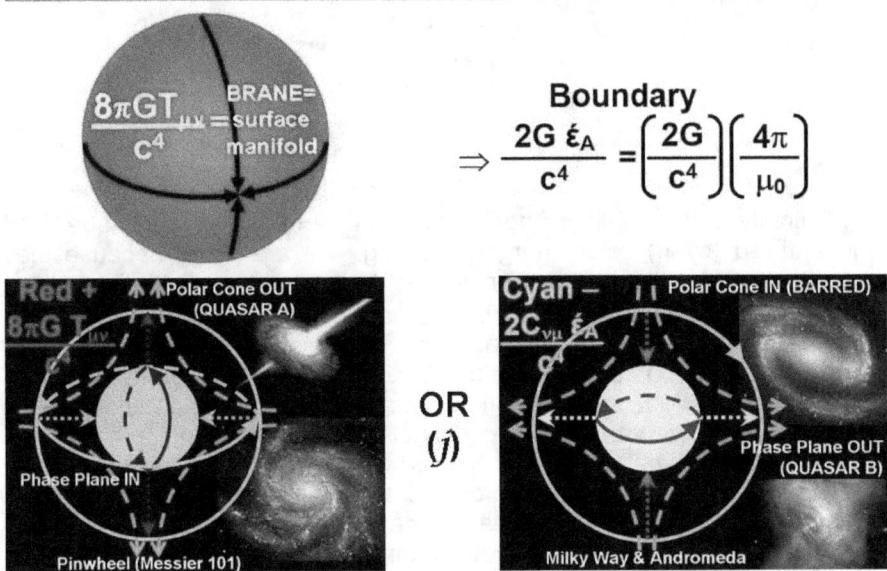

3.4: *Singularity/Brane Field Equation with Color Charge*

In this function, the value of a degenerate density ($\mu_0/2G$) is the spacetime container. Degeneration is covered in the last section of book. This density is satisfied by including a self-canceling angular permeability function: $U_A = \varepsilon_A/4 \times 10^{17} = 1$ kg/m s²:

$$\rho_m = [U_A/k^2] [\mu_0/2Gk^2 = 2.34378270633671 \times 10^{17} \text{ s}^2/\text{m}^2]$$

On pg. 26 we showed how to adapt the GFE to accommodate linear (red) or angular (cyan) singularity types. We can adapt Einstein's Brane Field Equation similarly. For singularity creation, the parent is developing the part of the function held constant. What is held constant is the enfolding brane. This brane's value is developed from radiant value such that it is the real part of the singularity.

The variable in the BFE defines the magnitude of the resulting brane. Again, this is the intrinsic enfolding brane. It is the baffling part of the singularity that confuses everyone. How this brane is defined determines how the universe shapes it into interactive phase. We see this shape as a something of a vortex.

Helicity v. Chirality

Helicity divides matter into right handed particles that move the same direction as they spin, and left handed that move the opposite direction of their spin.[31] These hands correspond with Fleming's rules.[32] This is the product of how the microstates in a change function define their spaces.

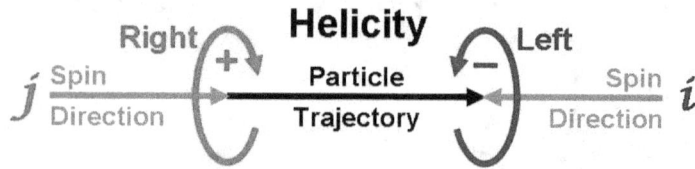

3.5: Helicity

Chirality is the quality of not being identical to its "mirror" image. Red and anti-red (cyan) color charges (see pg. 47) are chiral because they have the same right-handed change function (j), but their field dispositions are inverted. Such chirality describes virtual particle and anti-particle.

Photons (and presumably neutrinos) consist of entangled pairs of particle and anti-particle, reverting their chirality to two helicity states (± circular polarization, or = becomes linear).[31] Helicity of Weyl fermions prevents annihilation as chirals but cancels weak interaction and mass.[33]

A particle anti-particle pair is created by adding energy to the confined spaces of an entanglement band or "flux tube" pair until they quantize.[34] A photon from electron quantum leap is the most familiar example.[35] These always occur in positive and negative pairs, and not necessarily entangled as with photons. Gluons split into new pairs.

The energy going into this is light. Light has both attributed (EMA) and emitted (EMR) values in it.[36] Both EMA and EMR will contribute to

---

[31] Dreiling, J. & Gay, T. (2014). Chiral Photons and Electrons. University of Nebraska. http://physics.unl.edu/~tgay/content/CPE.html.
[32] Klauber, R.D. (2017). Chirality vs. Helicity Chart. quantumfieldtheory.info/ Chiralityvshelicitychart.htm.
[33] Romão, J.C. & Silva, J.P. (May 29, 2016). Helicity and Chirality. Instituto Superior Técnico. porthos.tecnico.ulisboa.pt/CTQFT/files/HelicityAndChirality.pdf.
[34] Smilga, A. (2001). Lectures on Quantum Chromodynamics. World Scientific.
Muta, T. (2009). Foundations of Quantum Chromodynamics: an Introduction to Perturbative Methods in Gauge Theories. 3 ed. World Scientific.
[35] Schombert, J. (2015). Quantum Physics. University of Oregon Department of Physics. abyss.uoregon.edu/~js/cosmo/lectures/lec08.html.
[36] Mattson, B. (Mar. 2013). The Electromagnetic Spectrum. imagine.gsfc.nasa.gov/ science/toolbox/emspectrum1.html.

momentum and new particle creation. When EMR quantizes, the perfect child splits into chiral particle and anti-particle. Like our singularities, EMA and EMR are applied in separate fields. Singularities have ideal EMA-EMR information sequence.

The division follows an EMA information pattern specific to the host's intrinsic values and microstates. We observe this in light,[37] which is subject to the same extratemporal change conditions as it bleeds from a source including a photon. Naturally this robs the system of all the associated momentum and related qualities.

The universe uses things that aren't there, conserves them because they quantize, AND consequently makes quantum wholes far greater than the sum of their actual parts. Antithetical to Aristotle's "A system is more than the sum of its parts."

The vortex shape is the environment acting ON the singularity, and limiting how things can interact with the singularity. Both diagrams are populated by colored lines and arrows. The brane space has its own concept of rotating into phase. A secondary field is rotating at a right angle to the first. Phase has its own rotation indicated by the grey circle with arrow on the outside of each diagram. The patterns of change functions are in the next chapter.

The type of singularity or degenerate being created can be ascertained looking at the surface features of the parent. The parent will have the phase features as an identity. The "child" is the antithesis of the parent: order goes in, disorder radiates out. The next two chapters show how color charges and change functions work and relate to light.

Color charge was not originally designed to reflect the actual color spectrum. Due to unit scale differences, linear radiant emission tends toward red (heat) and angular toward cold blue (UV). To form a red color charge field, surface emission would be oppositely blue. To form the cyan color charge, surface emission would be oppositely red.

Yellow, like our sun, is the subtractive form of blue. On the spectrum, however, it is intermediary between red and blue. It is not offering an extreme enough ordering to create a substantial singularity. On the same scale as a red or blue star needed to form singularity, it could produce a neutron star. This could then be captured to form a Thorne–Żytkow object[38] and potentially evolve into a singularity.

Referencing the diagram (pg. 47) again, Type A (red) ejects by polar jets to vacate its $T_{\mu\nu}$ "surface" retaining its volume-equivalent ground state value (its identity). This body can consume and emit simultaneously,

---

[37] Wood, D. (2018). Atomic Spectrum: Definition, Absorption & Emission. study.com/academy/lesson/atomic-spectrum-definition-absorption-emission.html.

[38] Levesque, E. & Massey, P. (June 4, 2014). Astronomers discover first Thorne-Żytkow object, a bizarre type of hybrid star. colorado.edu/today/2014/06/04/astronomers-discover-first-thorne-zytkow-object-bizarre-type-hybrid-star.

Quantum Relativity

fluctuating the radial absorption/emission space size holding density constant. It can be rotating quickly like a lighthouse as with pulsars. On a galactic level, these are pinwheel-style spirals.

Type B (cyan) ejects in circular pulses to vacate its "volume" $C_{\mu\nu}$ retaining its surface-equivalent ground state value (its identity). Again, radial volume increases with absorption then decreases with emission holding density constant. An electron, as an example, fills into another orbital level then bounces (quantum leaps) back releasing a photon.

The Crab Nebula pulsar is used as an example because there are time-elapsed videos of it actually doing this.[39] A sombrero galaxy is a good example, except the time elapse is a tad long. This type of black hole creates a bar-style spiral galaxy like the Milky Way. This is the shape of Lemaître's big bang.

There is also a third type where these two interact and try to annihilate each other. Of course there are also mergers, which are also quite fun with showing light obviously exhibiting itself out of character as a giant longitudinal wave (ref. LIGO[40]). Becoming supermassive requires merging and growing without dying from neglect.

A singularity is simultaneously completed by its light horizon—in total the bubble. This perturbation is Noether's conservation manifesting as a quantum fluctuation. It already has its perfect proportions. The disorder conditions "pop" it into existence and required are to continue its existence. This popping in and out of existence is quantum fluctuation.[41]

The universe conserves change by applying or removing it where it conveniently can. This is seen as quantum particles popping in and out of existence quantum fluctuation is a perturbation, meaning all the requisite values come into focus to define a thing. To sustain a virtual identity, the focal conditions must be maintained.

QR Field Shapes

In Relativity, spacetime does things. The quantum universe is mostly synonymous with pure mathematics. In mathematics, space is an ordering set with structure potential. Ordering requires a change function. You can think of a blank page as a space to add axes, scales, etc. Space specifically isn't doing anything until it gets defined.

Defining a space requires magnitude and change—which gives direction/order. These features technically make a vector space. All spaces in the quantum universe are vector spaces on some level, so we won't

---

[39] (May 20, 2013). Chandra Monitors the Flaring Crab. NASA Video. https://www.youtube.com/watch?v=PiXOaGuhlLQ.
[40] (Feb. 11, 2016). Gravitational Waves Detected 100 Years After Einstein's Prediction. https://www.ligo.caltech.edu/news/ligo20160211.
[41] Strassler, M. (Aug. 29, 2013). Quantum Fluctuations and their Energy. https://profmattstrassler.com/articles-and-posts/particle-physics-basics/quantum-fluctuations-and-their-energy/.

quibble. Instead we will redefine this **ordering set with potential for structure** as a **manifold**.

A manifold has two shapes: applied anomaly and smoothed. Smoothing is a role of the specific change function (see inset below). A manifold is a confined quantum variable, meaning it only occurs in a proportion with other variables and it can quantize. To create a manifold from scratch (detailed better in Matter section, notably pg. 119 et seq.):
1. Value accumulates in an existing available change space—typically an entanglement band/flux tube.
2. Accumulation activates the otherwise inactive available space.
3. That value quantizes—becomes a proportional unit relative to the change space provided.
4. Quantization triggers independent microstate sequence causing a change in relationship with the host (parent).

Redefining Entropy

We tend to think of change as transformation, conversion into another function (e.g. Leibniz) or context (e.g. Newtonian surface to volume). Transformations affect position or superficial qualities, but not the defining shape. This includes: rotation (orientation), translation (moving), reflection (flipping), observational dilation (expand/contract) and scaling (resizing).[42]

In modern Thermodynamics, entropy is used to indicate disorder—the inability to convert energy into work. Why not just say disorder? Disorder already is one readable word. The ambiguity of terms better fits a more mathematically complex role as it was originally intended for.

Entropy derives from ἐν (in) + τροπή (transformation, a turning, change), with Clausius meaning to transform between heat and work.[43] The first law ($dU=\delta Q-\delta W$) is the defining transform of Thermodynamics. Disorder (S-entropy) is the directional predisposition of that change (second law).

We will call these defining change actors entropies, specifying S-entropy (a different type of change) for clarity. As logical quantum functions, they are all treated primarily as unit quantum numbers. These quantum numbers associate directly with the quantum numbers known as colors or color charges, with anti-colors as their chiral forms.

The entropy of a quantum number identity like a singularity or color charge defines two manifolds. The manifolds of nu matter (singularities ν=order) are sequential, whereas the manifolds of mu (μ=disorder) are

---

[42] Stapel, E. (2016). Function Translations. purplemath.com/modules/fcntrans.htm.
[43] Carnot, S. (Aug. 5, 2014). Entropy. eoht.info/page/Entropy+%28etymology%29.

concurrent. The first manifold for each is given linear values, whereas the second receives angular.

Perspective determines how EMR (enfolded ε-space) and EMA (permeability έ-space) values apply to singularity manifolds. Each change function has a dominant feature appearing as the first sign in its definition. This is the feature whose boundary condition satisfied singularity (bold). The information sequence of both must be satisfied.

Red j = + OR − (**EMR** volume | <u>**EMA**</u> surface)
Cyan j = − OR + (**EMR** surface | <u>**EMA**</u> volume)

The second variable is the total absorption capacity—permeability (έ). Only matter can be absorbed, which is a hazard for singularities too small to absorb matter. The EMR-EMA information of the absorption is subject to Wheeler interaction (see pg. 172). Light goes around the singularity, even the increased volume not enfolded. Mass only matters to the ability to interact resulting in absorption. Mass otherwise makes no difference to a strongly interacting body.

The most significant feature is the composite full force of the singularity defining the polar/equatorial jets and intake. This is Pythagorean ($F^2=R^2+A^2$) and violates spacetime constraints. The universe cheats but without breaking rules. It widens toroidal aperture features (jets) and sets them away from the source far enough that no violation occurs. These affect local flux (flow) in and out.

If we ignore the rest of the universe around the singularity, we already have one of two sorts of vortex whirlpools. Einstein's visualization of this was truly brilliant, and we must applaud his effort to break it down and explain it. He could never conceive most of this is the universe acting on the singularity. Without that, the singularity dies.

3.6: ESA/Hubble "Dark Matter" Kaleidoscope

When a singularity dies, it stops defining the spacetime fabric of its bubble. The effects are still there, the singularity just isn't there to keep them under control. They now begin to respond to other singularity-bubble fields. Their densities are different, especially nearest to where the singularity was. The result is a kaleidoscope effect. Like the Hubble picture on the cover or above[44].

Dark matter is our generic term for matter that does not reflect or emit light. Unless we see an interaction, we honestly can't say what it is. When light passes through that space and gets distorted, we blame it on dark matter. Fact is, the dark matter is likely gone and we are simply observing the left over distortion of it having been there. It is now subject to being smudged around by active fields.

As we just noticed, this dark matter isn't actually there. It was. It somehow got twisted out of the Occam's Cycle—most likely by neglect. This leaves the kaleidoscope effect adding a Möbian-like twist to a deeper field. The twist on the cover is between us and the cluster MACS J0416.1-2403 at z=0.397.[45] That is practically next door when CN-z11 is at z=11.1.

Vortices

Vortices are described as "whirling" and spiral-shaped; generally in terms of matter whirling around as in cyclones, tornados, hurricanes, etc. This term is often used to describe quantized angular momentum (QAM)—a state of particles rotating in a circular motion around a common origin.[46]

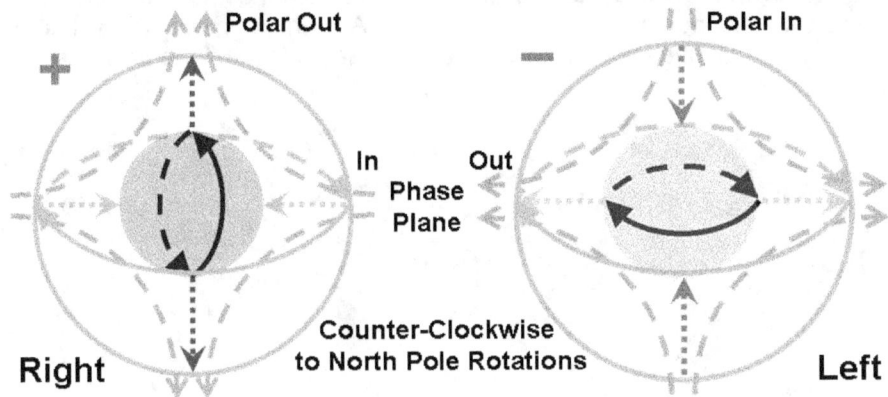

3.7: Phase on Red/Cyan Singularities

---

[44] (Jul. 21, 2016). Space... the final frontier. ESA/Hubble Information Centre. https://phys.org/news/2016-07-space-frontier.html.
[45] Jauzac, M. et al. (Oct. 22, 2014). Hubble Frontier Fields: The Geometry and Dynamics of the Massive Galaxy Cluster Merger MACSJ0416.1-2403. https://arxiv.org/pdf/1406.3011.pdf.
[46] Nave, C.R. (2017). Quantized Energy States. Georgia State University. http://hyperphysics.phy-astr.gsu.edu/hbase/Bohr.html.

Quantum Relativity

Vortex conditions can be static (strictly circular) or dynamic (degrading toward the center or evolving away from the center). The term vortex best describes a dynamic system.[47] Singularities are relatively static. The dynamics are always the environment acting on the inert singularity. We assume a temporal phase space vortex around a black hole.

From a cosmic perspective, space and time are flat. As we will see over the next chapters, flat spacetime is constructed from change interactions. Applying a quantized change function to a segment of flat spacetime cause differentiation of the change functions—they unwind.

Flat spacetime easily defines a volume in two or three axes. A singularity creates a volume with one axis. Its presence takes that axis out of the local phase definition leaving a sinusoidal change dynamic flattened in time. As helical time unravels, the orbital twists of matter around the singularity exaggerate.

The result is a context-sensitive vortex around the singularity that respects its polar and equatorial strengths. The rotational rate of the vortex is determined by the relationship between the matter of that phase space and the singularity it is nucleated to. Temporal dilation is causal of this nucleation.

Such a rotation can be described as quantized angular momentum (QAM). We classify orbital behaviors among general electromagnetic fields emerging from degenerate identities (see pg. 162 et seq.). There are twisting features in the orbital field equations, even dating back to Newton. This gives the QAM a mechanical effect maintaining the perturbation of singularity. A singularity is confined in its QAM environment. We will soon see that singularities, although inert, are the generators of the universe.

---

[47] Ting, L. & Klein, R. (1991). <u>Viscous Vortical Flows</u>. Springer Berlin Heidelberg.

Quantum Field Theory

Spacetime tells matter how to move; matter tells spacetime how to curve.

—John Archibald Wheeler on gravity, 1998
<u>Geons, Black Holes, and Quantum Foam</u>

Quantum Relativity

Change-Color Truth Tables

Entropies to Colors

×	j	$\hbar$	i	$\hat{j}$	$\hat{\hbar}$	$\hat{i}$
j	−1	$\hat{i}$	$\hat{\hbar}$	0	i	$\hat{\hbar}^2$
$\hbar$	$\hat{i}$	−1	$\hat{j}$	i	0	$\hat{j}$
i	$\hat{\hbar}$	$\hat{j}$	−1	$\hat{\hbar}^2$	j	0
$\hat{j}$	0	i	$\hat{\hbar}^2$	+1	$\hat{i}$	$\hat{\hbar}$
$\hat{\hbar}$	i	0	j	$\hat{i}$	+1	j
$\hat{i}$	$\hat{\hbar}^2$	$\hat{j}$	0	$\hat{\hbar}$	j	+1

×	r	g	b	c	m	y
r	−1	y	m	0	b	m
g	y	−1	c	y	0	c
b	m	c	−1	m	r	0
c	0	y	m	+1	b	g
m	b	0	r	b	+1	r
y	m	c	0	g	r	+1

Permittivity & Permeability Constants

Legend:

- ε = Permittivity (energy application/potential)
- $\acute{\varepsilon}$ & μ = Permeability (energy flow)

- 0 = Maxwell's magnetic constants
- L = Newtonian linear/gravity constants*
- A = Coulomb's angular constants
- S = Planck's spin constants
- V = Vector axis relational constant (2 sec φ)*

Permittivity

Symbol	Constant	SU
$\varepsilon_0 = 1/\mu_0 c^2$	3.54167512704815E−25	kg/m³
$\varepsilon_L = \varepsilon_S/\varepsilon_A \varepsilon_V$	1.34103117585718E+27	kg/m
$\varepsilon_A = 1/4\pi\varepsilon_0 = k(\varepsilon)$	2.24688794684204E+23	m³/kg
$\varepsilon_S = c^2/\acute{\varepsilon}_S$	8.52246609859054E+50	/kg s
$\varepsilon_V = \varepsilon_S/\varepsilon_L \varepsilon_A = 2\sqrt{2}$	2.82842712474619E+00	/m² kg s

Permeability

Symbol	Constant	SU
μ_0	3.14159265358979E+07	m s²/kg
$\acute{\varepsilon}_L = G = c^2/\varepsilon_L$	6.70197080364171E−11	m³/kg s²
$\acute{\varepsilon}_A = c^2/\varepsilon_A = 4\pi/\mu_0$	4.00000000000000E−07	kg/m s²
$\acute{\varepsilon}_S = \hbar$	1.05457172646947E−34	kg m²/s
$\acute{\varepsilon}_V = \acute{\varepsilon}_S/\acute{\varepsilon}_L \acute{\varepsilon}_A = 1/\varepsilon_V c^2$	3.93381199861553E−18	kg s³

* Explanation on pg. 76 et seq.

4. Change Functions

Einstein used new variables and language to evolve the field concepts of Poisson and Gauss into the more contemporary thinking of Riemann and Ricci-Curbastro.[1] His evolution helped solve the old thinking problems, and created its own list of unforeseen new problems.

Quantum Mathematical Language (QML) helps us evolve to the next level of problems. It explains the architecture and its field theory which solve these new problems. Change functions increment, sequence, and shape everything. They help us construct spacetime from scratch, revealing confined (hidden) quantum perspectives.

Each change function has two distinct fields: one order with explicit value and the other disorder with attributed value—relating them to the entropy of Thermodynamics. For this and the classical meaning we call them entropies. They are vital operators giving action to mathematics, often overlooked in plain view. Among the ranks of functions, these operators define cycles placing them at the root of the architecture.

Entropies

Entropies use Boolean logic OR ($j=+|-\equiv$cosine), AND ($i=-\&+\equiv$sine), BOTH ($h=j\&i\equiv$cotangent) | NOT (t=real positive) truth concepts.[2] In each of these sequences, the first variable is the explicit order value.

Each relates to an axis and is governed by rotational rules such that they are solutions to $\sqrt{-1}$, and chiral forms to $\sqrt{+1}$. Except time. Time is antithetical to hypercomplex making it a linear and positive real value.

Complex functions contain on imaginary element. Hypercomplex functions contain two imaginary elements. We prefer the word **virtual** over complex because it conveys better into forming and evolving matter.

As with the relationship of light and lens with an image, virtual describes a divergence.[3] Likewise with matter, real consists of a stable convergence where virtual is unstable subject to divergence. Careful choice of words enables scalable application.

Leonhard Euler (1707–83) developed the familiar imaginary number ($i=\sqrt{-1}$) into a system of complex variables to relate algebraic and polar axes (constructing ordinary spaces).[4] He did this based on context without

---

[1] Vasconcellos, C.A.Z. (editor). (2016). Centennial of General Relativity: A Celebration. New Jersey: World Scientific.

[2] Zimmer, S. (Jun. 30, 2017). What is a Boolean Operator? Alliant International University Library. library.alliant.edu/screens/boolean.pdf.

[3] Pumplin, J. (2000). Images Real and Virtual. web.pa.msu.edu/courses/2000fall/PHY232/lectures/lenses/images.html

Quantum Relativity

any evidence of a logical explanation of HOW i^2 becomes -1. We will use similar illustrative methods with graphical applications later.

Tessarine and quaternion truth tables devised later (below) have a mutual problem: contextual validity with no tangible explanation HOW. Quaternions are asymmetric, making sequence important so ji=–ij. In all three, ij=k or –k.[5] Cockle's tessarines are symmetric (ij=ji) and use j^2=+1.[6] Contemporary tessarines are j^2=–1.[7]

Asymmetric					Symmetric								
Quaternions					Tessarines$_O$					Tessarines$_N$			
×	j	k	i		×	j	k	i		×	j	k	i
j	–1	i	–k		j	+1	i	k		j	–1	i	–k
k	–i	–1	j		k	i	–1	–j		k	i	+1	j
i	k	–j	–1		i	k	–j	–1		i	–k	j	–1
Hamilton					Cockle					Negulescu			

4.1: Quaternions v. Tessarines

Around 1847, George Boole (1815–64) devised the system of syllogistic logical operator concepts: OR, AND, and BOTH/NOT.[8] NOT (t) is excluded here. Although ideal for the task, this logic was never used to explain HOW operators work in truth arguments like tessarines and quaternions.

Boolean		Symbolic Logic			Entropy	Topology	
AND	x & y	x ∧ y	Kxy	XAND	i	tetrahedron	z
OR	x | y	x ∨ y	Jxy	XOR	j	circle	x
NOT	\ x	¬ x	Ix | Nx	NOTX	t	flat	t
BOTH	x ‖ y				h	sphere	y

4.2: Boolean Logic Symbols

---

[4] Sachs, R. (2011). Euler's Formula for Complex Exponentials. math.gmu.edu/~rsachs/m116/eulerformula.pdf.
[5] Krishnaswami, G.S. & Sachdev, S. (Jun. 2016). Algebra and Geometry of Hamilton's Quaternions. ias.ac.in/article/fulltext/reso/021/06/0529-0544.
[6] Cockle, J. (1848). On Certain Functions Resembling Quaternions and on a New Imaginary in Algebra. Philosophical Magazine and Journal of Science. pg 436. biodiversitylibrary.org/item/20157#page/450/mode/1up.
[7] Negulescu, V.L. (May 2, 2015) Hyper-Complex Numbers in Physics. http://article.sapub.org/10.5923.j.ijtmp.20150502.03.html.
[8] Norman, J. (Jan. 23, 2018). George Boole Develops Boolean Algebra. www.historyofinformation.com/expanded.php?id=565.

Quantum Relativity

This was resolve by accident of technological necessity substituting j for ±. By making ± a unit variable operator, it acquired the logical qualities of Boole's OR, inverting roles (flipping signs) every operation. AND was applied to i, but only inverts in multiples (as with h). To accommodate BOTH as a third solution ($h^2 = ij = -1$), negative comes first $i = -\&+$.

Among tessarines and quaternions, k=ij or −k=ij. This is an extremely limiting approach. We need the root $\pm k \equiv h^2 \equiv ij$. The hypercomplex variable can be reached other ways as well, such as $2h^2 = i^2 + j^2 \equiv 2(jx + iy)/z$, etc. Our entropy operators are not just features of a truth table. They define truth and are multi-functional.

Into Color

Having two more complex operators proved incredibly helpful in a broad range of applications. Everything about these operators is extremely ambiguous, so errors are easy. Fortunately, colors offer a real world analog. Most of how we describe, interpret, and otherwise define reality is fallible. It all goes out the window as soon as our understanding increases. Empirically established facts, however, never change.

Colors are more than just a metaphor for strong interactions. Handled properly, they show us everything we need to know about particle interactions and hadronization. They were selected for QCD due to the qualities they exhibit on the artist's pallet when mixed together. For a quantum truth table, this is a watershed victory because we can show it in the ordinary world and let nature resolve the ambiguities.

COLORS			ANTI-COLORS			
Red	r	j	Anti-red	$\bar{r}=c$	j'	Cyan
Green	g	i	Anti-green	$\bar{g}=m$	i'	Magenta
Blue	b	h	Anti-blue	$\bar{b}=y$	h'	Yellow
White	rgb	K	Black	cmy	K	

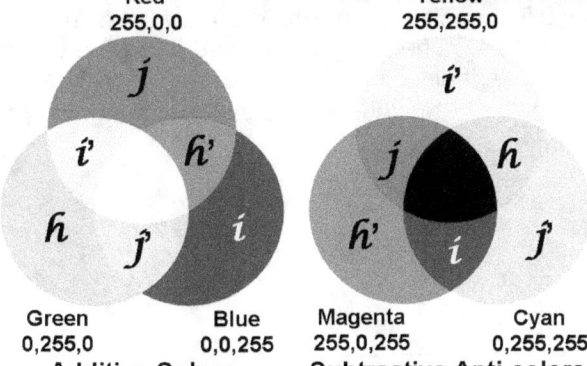

4.3: *Color to Change/Charge & Palette*

Quantum Relativity

Chromodynamics normally uses colors and anti-colors with the same symbols but with a bar across the top. We will use the subtractive cmy=K colors as anti-colors for purposes of visualization. You do still have to remember the r-c, g-m, and b-y color anti-color relationships.

Color palates show how colors and anti-colors mix with their own kind to produce their opposites.[9] If two colors interact, the nature of their interaction is provided by the anti-color of their intersection. Degree of intersection distinguishes different types of interactions. As such, the color truth table is vital to how all particles are put together.

For readability, additive color only is used in the text. Negative and inversion cause different orientation changes. Orientation is easily confused with the chiral anti-function. The chiral forms are thus shown with a prime: j', v' and μ', etc.

Entropies

×	j	$\hbar$	i	$\hat{j}$	$\hat{\hbar}$	$\hat{i}$
j	-1	$\hat{i}$	$\hat{\hbar}$	0	i	$\hat{\hbar}^2$
$\hbar$	$\hat{i}$	-1	$\hat{j}$	$\hat{i}$	0	$\hat{j}$
i	$\hat{\hbar}$	$\hat{j}$	-1	$\hat{\hbar}^2$	j	0
$\hat{j}$	0	$\hat{i}$	$\hat{\hbar}^2$	+1	i	$\hat{\hbar}$
$\hat{\hbar}$	i	0	j	i	+1	j
$\hat{i}$	$\hat{\hbar}^2$	$\hat{j}$	0	$\hat{\hbar}$	j	+1

Colors

×	r	g	b	c	m	y
r	-1	y	m	0	b	m
g	y	-1	c	y	0	c
b	m	c	-1	m	r	0
c	0	y	m	+1	b	g
m	b	0	r	b	+1	r
y	m	c	0	g	r	+1

4.4: Color/Change Truth Tables

These two tables are symmetrical: $ij=ji$ (or br=rb). They say exactly the same thing in two different ways. The first uses the entropy containers represented by the mathematical operators. The second uses the colors of chromodynamics. For comparison, all the truth tables in this section are arranged by palate rgb ($j\hbar i$) and cmy ($j'\hbar' i'$).

These tables are much bigger than their predecessors because we have three specific conditions and their chiral forms. In the predecessors we see negative like $-j$. The analysis showed there are two possible chiral conditions for this variable. In one condition, j is simply rotated and looks like j' except it isn't j'. Rotating alone does not change the identity, hence imperfect mirroring.

---

[9] Brown, A. & Feringa, W. (2003). Colour Basics for GIS Users. Harlowe, England: Pearson Education Limited.

Venn Algebra

Venn diagrams[10] are taught in various logic courses like algorithms and symbolic logic. They offer a mechanical way to visualize the concepts of sets, including subset, superset, and the interactions of sets. For now, let us see how our operators function in an applied mathematical way by Venn-like.

$i+i =$ (−/+) + (−/+) $= 2i$ $i^2 =$ (−+/+) $=-1$ $-i=$ (+/−) ≡ (+/−) $= i$

$i'+i' =$ (+/−) + (−/+) $= 0$ $i'^2 =$ (+/+ −/−) $=+1$ $h =$ (+/− −/+) $=$

$j+j =$ (−/+) + (+/−) $= 0$ $j^2 =$ (+/− −/+) $=-1$ $-j=$ (−/+) ≡ (−/+) $= j'$

$j'+j' =$ (−/+) + (−/+) $= 2j'$ $j'^2 =$ (−/+ −/+) $=+1$ $\boxed{h'^2 =\ (−/+)(−/+) =+1}$

4.5: Color/Change Operation Diagrams

Aside from the logical AND, OR, and BOTH features, the operators will invert their signs contextually. Color and anti-color entropies have inverted rules. Where j inverts sign with each operation, j' never inverts (see rule table pg. 63). This is easiest to see in the counter-intuitive $k+k=0$ of j, h', and i'. The opposite for these is also true: $k-k=2k$.

Hypercomplex h is also known as phase entropy. The simplest definition is $h^2=ij$. While simple, this hides the fact that i and j of make up h by being irrational: $h^2=-i/j$. This causes the axis positions of i and j to align as needed to form h. It also rationalizes and provides anti-phase: $h'^2=ij'=-ij=j'/i'$ or

$$(-|+)/(+\&-)=(-|+)(-\&+)=(+\&|+)$$
showing $j\equiv 1/j'$; $-h\equiv h'$; $-i\equiv i'$.

For basic algebra, as with chromodynamics, these equivalences can be treated as equal quantum numbers (units). The issue we have with $\pm k=ij$ is relative magnitude. Relative to $j=1$, $i=e$. This creates red-blue and cyan-yellow bonding limitations for QCD. To form h practically of one magnitude, i of a lesser magnitude combines with j of the same magnitude.

Phase ($h=e$ to $j\equiv i=1$) is a higher class of change function. It sees unit equivalence allowing a single bond with one (rg, bg, cm, my=Type I Weyl fermions) or both (rgb or cmy=Type II Weyl fermions). Both can only be done simultaneously. We will see later how these combinations define volumes by axial rotations (pg. 87 et seq.).

---

[10] Stapel, E. (2017). *Venn Diagrams*. purplemath.com/modules/venndiag.htm.

Quantum Relativity

The change operator of order is j. The change operator of **disorder** and **distribution** is i. At the same magnitude their proportions are $1:e$. To interact into phase h, they must be of the same scale (1:1) requiring i to be the next lesser magnitude. This is why fusion discharges so much energy. From their interaction changes emerges the quality of time resisting that interaction.

If we have a mixture of i and j (simple addition=+), the two can interact and pass through the same space. Separately, their axes define surfaces. In degrees of interaction (inner sum), a transient (weak) field cannot distinguish surface from volume and applies value to both. Full bonding interaction results in a **flavor** volume.

Evolving Venn Diagrams

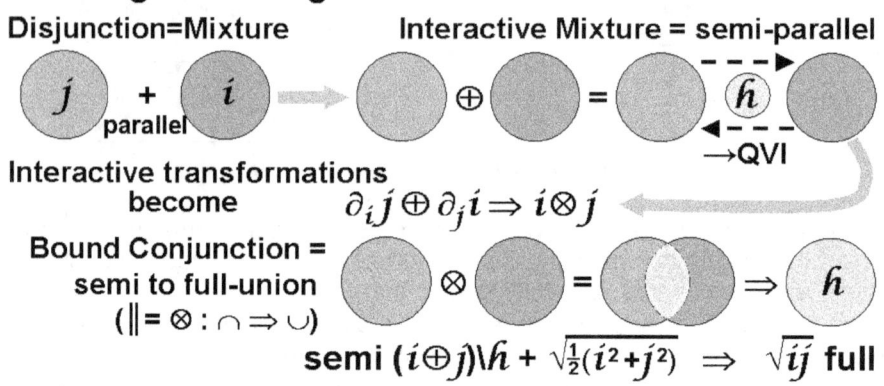

4.6: Evolving Venn Diagrams (to Hypercomplex)

The partial derivatives (∂) are j and i transforming into each other. They are smoothing into a common state of proportional (1:e) equilibrium. This is the change root of manifolds smoothing from an anomalous atlas (point distribution[11]). Vector energy follows. The lead is scalar energy adjusting value to equilibrium (2nd Law of Thermodynamics[12]).

The prevailing/dominant entropy condition of a set (e.g. the super set conditions) is its phase identity. Changes to phase identity affect the parts (quantum fluctuation). This is shown as an inner product function. The inner product is like the inner sum. Only parts are involved, while the rest is excluded.

Normally, we apply operators assuming all of each variable participates, like 6×4=24 and 4+5=9. The algebra of Venn logic is ambiguous. Just because you have 6 of one and 4 of another doesn't mean you are going to use all those parts, or necessarily use them in the

---

[11] Welbourne, E. (Mar. 18, 2017). <u>Atlases and Charts of Smooth Manifolds</u>. http://www.chaos.org.uk/~eddy/math/smooth/atlas.html.
[12] Redner, S. (2006). <u>Equilibrium and the Second Law of Thermodynamics</u>. http://physics.bu.edu/~redner/211-sp06/class-macro-micro/2nd-law.html.

same ways. It is like baking a cake: in theory you didn't use everything in the kitchen.

When we complete our evolving operation, the total scalar value of h is equal to that of i or j. This leaves half the scalar value of the original parts as excess: $(i+j)\backslash h$ reads the mixture of i and j not (\) in h. That NOT feature is a time variable making use of what is not used.

We can change the identities of quantum particles simply by changing their phase identity. Neutrino bands given energy rotate to free value making them relativistic. The information in that energy affects the intrinsic information, swapping phase identity. This is called "oscillation."[13]

Algebraic Logic

In algebra our operators are fairly simple: $(h|i|j)^2 = -1$ and $(h'|i'|j')^2 = +1$. Color entropies are imaginary ($\sqrt{-A^2} = kA$) and anti-colors are REAL ($\sqrt{A^2} = k'A$). The operators easily disappear into any unit. That means attaining or applying any quantum number, like a unit circle, permeability or permittivity, satisfies a unit operator. Sometimes it is simply hiding in the variables or in the temporal units. If an operator is present, it has profound contextual applications. In algebra, the operators are explicitly stated.

Primary	Boolean Operator	Secondary	Alternates		Applied				
			Sum	Product	k^2	$k+k$	$k-k$	Relative k^2	$(x+ky)^2$
h	i BOTH j		n	n	-1	$2h$	0	$ij = -i\backslash j \mid i'/j$	$x^2 + 2hxy - y^2$
i	$-$ AND $+$		n	y	-1	$2i$	0	$-hj = h/j' \mid j/h'$	$x^2 + 2ixy - y^2$
j	$+$ OR $-$		y	y	-1	0	$2j$	$-hi = h/i' \mid i/h'$	$y^2 - x^2$
h'	i' BOTH j'		y	y	$+1$	0	$2h'$	$-ij = j/i \mid j'/i$	$y^2 + z^2$
i'	$+$ AND $-$		y	n	$+1$	0	$2i'$	$hj = h'/j \mid j/h$	$x^2 + y^2$
j'	$-$ OR $+$		n	n	$+1$	$2j'$	0	$hi = h'/i \mid i'/h$	$x^2 + 2j'xy + y^2$

4.7: Change Logic Rules Table

All entropies are at least complex operators differing by Boolean function and when they alternate signs. The chiral forms have opposite sign alternating rules. For example, j and h' alternate signs each operation, but j' and h never alternate signs.

The table shows how these translate algebraically. Some of these qualities make finding the roots of certain functions, like $x^2 - y^2$ or $x^2 + y^2$ fairly easy. They can also be used to analyze quadratic functions. This assumes

---

[13] Casper, D. (1998). Neutrino Oscillations. ps.uci.edu/~superk/oscillation.html.

Quantum Relativity

the axes involved are functions of the entropies. Each entropy correlates to and affects a specific axis: $j \to x$, $i \to y$, and $h \to z$. However, a generic $\pm \to j$.

$$ax^2 + bx + c = 0 \quad \text{if } 4ac > b^2 \text{ then } k = h^2 \text{ else } k = h'^2$$

Multiply through by 4a
$$4a(ax^2+bx+c=0)$$
$$4a^2x^2 + 4abx + 4ac = 0$$

Isolate 4ac
$$4a^2x^2 + 4abx = -4ac$$

Add b^2 to both sides
$$4a^2x^2 + 4abx + b^2 = b^2 - 4ac$$

Take root and apply k
$$2ax + b = j\sqrt{k(b^2 - 4ac)} = u$$

For $4ac \le b^2$ = r-phase
$$= j\sqrt{(b^2 - 4ac)}$$

For $b^2 < 4ac$ = z-phase
$$= j\sqrt{(4ac - b^2)}$$

Giving a generic
$$x = \frac{u - b}{2a}$$

4.8: Quadratic j-Change Example

The generic foils and roots identities assume we apply the right axes. The most baffling and algebraically useful feature of j, h', and i' is their counter-intuitive zero sums (e.g. $j+j=0$) and doubled differences ($j-j=2j$). We could make the list considerably longer, but that would belabor the point.

1. $(A \pm B)^2 = A^2 \pm 2AB + B^2$
2. $(A - jB)^2 = A^2 + 2j AB - B^2$ always $(A + jB)^2 = A^2 - B^2$
3. $(A \pm iB)^2 = A^2 \pm 2iAB - B^2$ $(iA + jB)^2 = -A^2 - B^2$
4. $(A \pm hB)^2 = A^2 \pm 2hAB \mp B^2$
5. $(iA + jB)^2 = (hC)^2 \Rightarrow h^2 = \dfrac{A^2 + B^2}{C^2}$ never $(ABC)^2 = \dfrac{i^2 + j^2}{h^2}$
6. $2h^2 = ij - ij = i^2 + j^2$
7. $2j = j - j = -2ih^2 = -i(i^2 + j^2)$ note $0 = j + j$
8. $\sqrt{j} = h\sqrt{-i} \equiv \sqrt{-i}$ and $\sqrt{i} = \dfrac{\sqrt{-j}}{h} \equiv \sqrt{-j}$

$$i = (a + ib)^2 \text{ but } hz = jx + iy \Rightarrow \sqrt{i = \dfrac{x^2 + y^2}{jz^2}} \equiv \sqrt{-j}$$

4.9: Imaginary Foils/Roots

One of our greatest challenges is finding exactly which operator applies to what. For example, $iA + jB \equiv jA + iB$. In algebra, we are pretty much just going through practice moves, so very likely we don't care. Here we do, so we will get to what exactly these operators can apply to. For algebra, we need to understand how they relate to each other.

General Unit Form:

$$\hbar^2 = ij = \frac{i^2 + j^2}{2} = \frac{ij - ij}{\pm 2}$$

Pythagorean

Unit States

presumptive $\quad \dfrac{i}{-j} = -1 = \hbar^2 \quad$ Co-imaginary

Möbian $\quad \dfrac{j}{i} = +1 = \hbar'^2 \quad$ Co-real

Null/Zero $\quad \hbar^2 + \hbar'^2 = 0 = \hbar_0^2$

4.10: Hypercomplex Phase Logic

We can also use a quadratic approach:

Quadratic Rationalization of i in $(ax+ib)^2=0$

$Ax^2 + Bx + C = 0 \Rightarrow 2Ax = -B \pm \sqrt{(4AC - B^2)}$

$a^2x^2 + 2iabx - b^2 = 0 \Rightarrow \underbrace{2a^2x = -2iab \pm ab\sqrt{8}}_{ax = -ib \pm b\sqrt{2}} \Leftarrow \begin{cases} 4(-ba)^2 - (2iab)^2 \\ = 4(ba)^2 + 4(ab)^2 \\ = 8(ab)^2 \end{cases}$

$x = \dfrac{-ib \pm b\sqrt{2}}{a} \quad$ solved for x $\quad|\quad$ solved for i $\quad i = \dfrac{-ax \pm b\sqrt{2}}{b}$

Recursive x: $\quad x = \dfrac{ax \pm 2b\sqrt{2}}{a} \quad\quad$ substitute into i

$j = \dfrac{ib + ax}{b\sqrt{2}} = \pm 1$ where b=1 all a

4.11: Quadratic i-j Relationship Example

    It seems unreasonable to go into too much detail on the anti-entropies at an algebraic level. It is easy enough just following the rules, but perhaps more information than is reasonable without a lot of application. Often such applications are hiding in plain view. Problem is understanding them enough to interpret and make testable predictions.

Quantum Relativity

Polar Refresher

To plot a point (5,60° shown):
1. Assign length values to the circles.
2. Find the angle (straight lines intersecting the origin). Positive angles are counterclockwise, negative angles are clockwise.
3. Measure the distance from the origin along the straight line relative to the circular distance markers. Positive distance goes toward the angle, negative away from the angle.

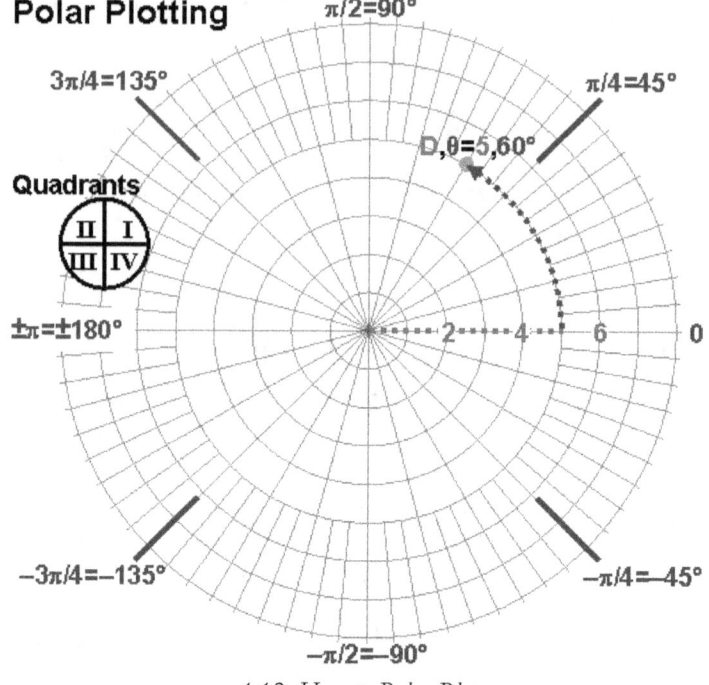

4.12: *How to Polar Plot*

The coordinate is represented as a single item by the Greek letter rho (ρ). If the above were ρ=−5,60°, then the point would be in the third quadrant and equal to 5,−120° or 5,240° depending on which direction you go. This is important because trigonometry functions are technically circular and should be plotted in polar coordinates.

Sine and cosine draw circles (below): sine in quadrants I & II, cosine in I & IV. These circles aren't just drawn once. They are drawn twice in the same places because of the negative values. If we used absolute values of these functions, we would see both circles for each.

The insets of the polar graphs for sine and cosine are the conventional/common way the functions are depicted. As we discussed in Color Geometry, these graphs are useful for measuring related magnitudes of spacetime density.

We also mentioned that the positive radius value (y-axis shown in green) need not have the same magnitude as the negative. Whatever the magnitudes of these are, they a quantified, meaning they have been reduced to a single unit.

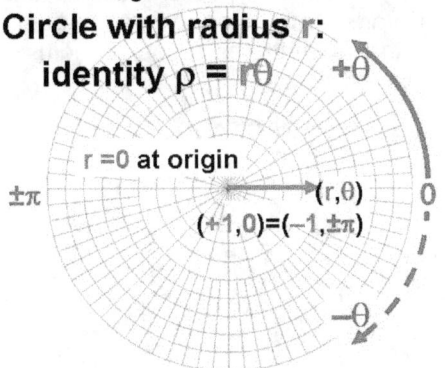
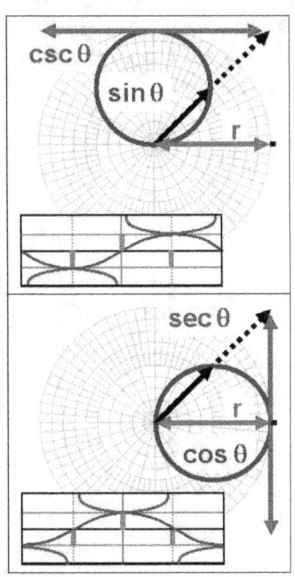

Circle with radius r:
identity $\rho = r\theta$

r =0 at origin

r on a polar graph is shown as "**y**"-axis for circular functions.
θ stretched into "**x**"-axis.
DO NOT confuse them with Cartesian
x=cos θ or **y**=sin θ.

4.13: Circular Function Polar Plotting

The x,y axes of the common depiction are confusing to students. They aren't actually x and y. The x-axis is the angle called the radian (shown in circumference increments of π). Without a radius, the length of this line is technically zero.

Because the radian marks are constant, we use a unit circle (radius=1) and fix the increments of the x-axis. The y axis value is the linear distance from the origin. As with other rectangular systems, you can assign value to the tick marks contextually.

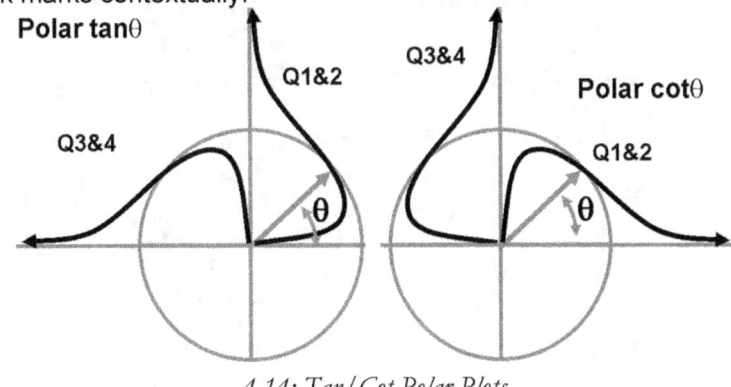

4.14: Tan/Cot Polar Plots

Quantum Relativity

Context Graphs

The graphs below show various ways to apply complex and hypercomplex operators (entropies) to different types of ordinary functions. The second column does things Euler's way: using xyz rectangular/linear axes. The first column recognizes that sometimes a scalar radius occurs instead of a linear axis, or with the linear axes.

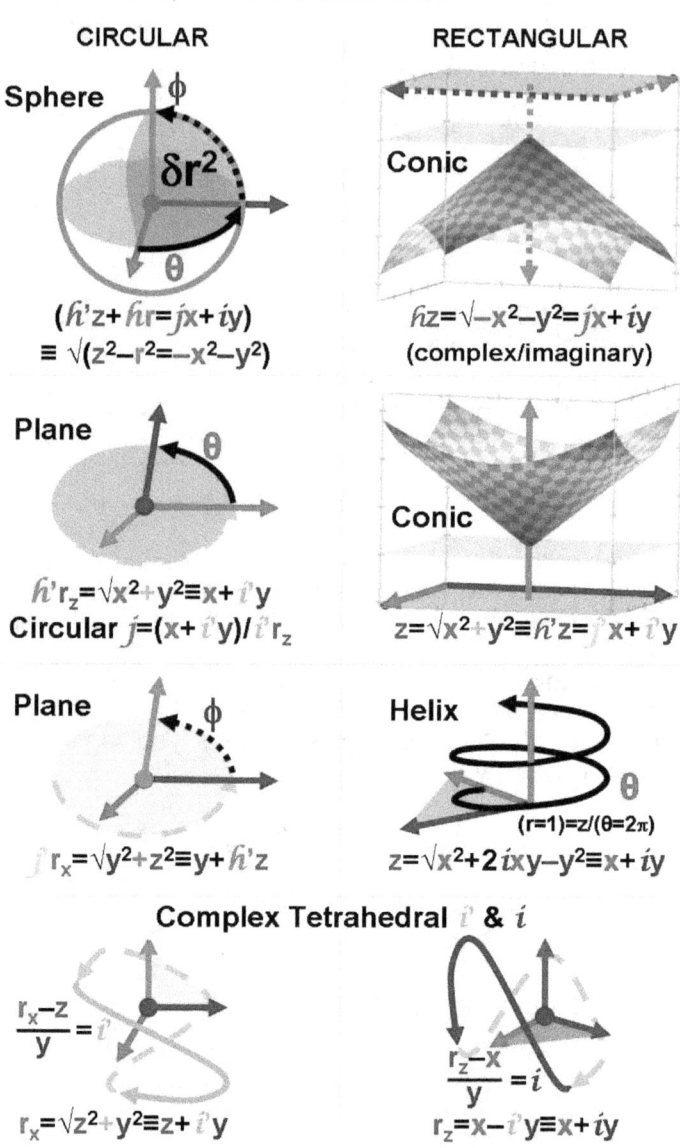

4.15: 3-D Algebraic Context Graphs of Changes

Quantum Relativity

Entropies also serve as radians due to change being cyclic. This is how circular planes form from j as flat (shown above as θ) and i as tetrahedral (ω=sinusoidal/S-shaped, lower right of above diagram). Both have the same radian length (2π) and combine to form a spherical radian (shown above as δ) in h of θ²+ω²=2δ→8π.

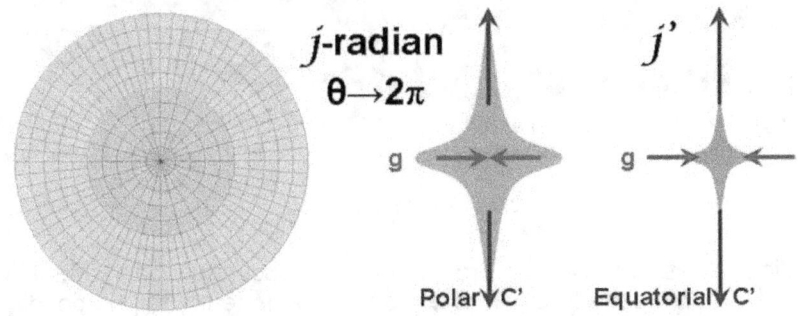

4.16: Flat Radian to Cross-Section Profile

In the polar graph above, red is gravity and the blue centrifugal plane is at a right angle to the red. The result in a cross-section profile appears complex. Gravity is a contracting spacetime (+) and centrifugal is expanding (−). Taken as unit values the conversion from one to the next is cosine.

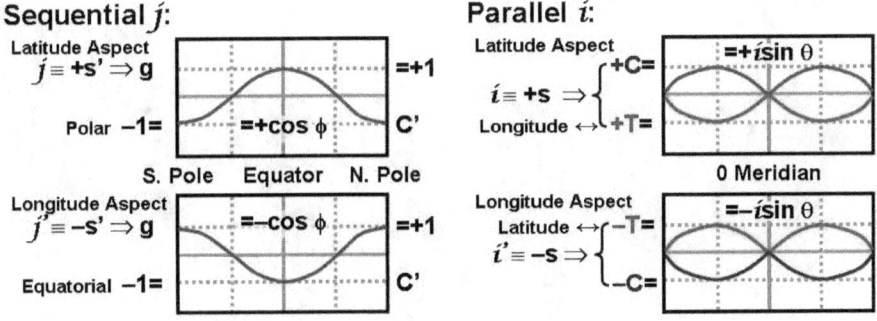

4.17: Applied Circular Functions to j&i Changes

Where j vectors are sequential, i vectors are simultaneous and working against each other (e.g. left handed). The i radian (ω) is only useful for showing these fields forming their plane. It is important to remember the vectors of i and j are linear and angular, which draws each of these spaces accordingly.

The i and j planes (below) are special purpose applying as ideal only in Abstract Phase. That covers every extra-temporal interaction, such as shaping Quantized Vector Interactions (QVI), which corresponds with

Quantum Relativity

mapping energy distribution through a microstate sequence, and their application to light.

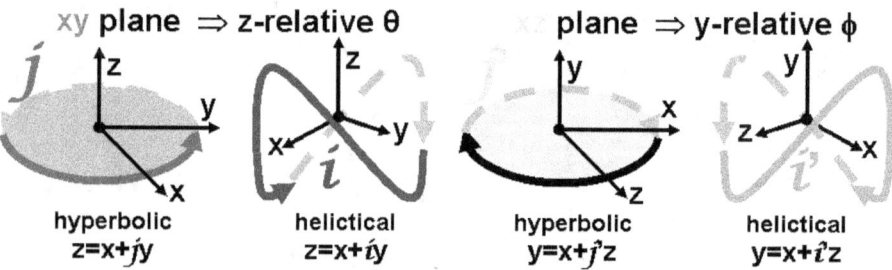

4.18: ĵ&î Change Planes

Light "sees" the plane in profile—or more accurately can't see it. To it there is an unimpeded space. The propagation is temporal, responding to the shaping of these surfaces coming into phase, and recognizing occupancy. This gives us both gravitational lensing and its related kaleidoscope effect.

5. Quantum Forces

Each change operator defines two fields: one contracts into order, the other expands to disorder; one is linear, the other angular, combining in a common spinor. The values are scalars giving magnitude specific to their context such as a space.

Order and disorder can also be described as generator and engine, relating this breakdown to Fleming's rules. The scalar energies are fundamental magnitudes, familiarly known as force and by wave propagations. In field interaction, forces act by means of permeation—applying value to spaces. The most popular way to convey scalar energy is by particle carrier due to focus (order).

Without focus of a change function, force propagates as disorder (light). Disorder is shaped by i into a transverse wave without spatial or temporal definitions of its own (light). Observation of quantum force uses virtual material reference.

In focus, forces form virtual particles (order). Virtual means quantum (uncertain) and transient (perturbative).[1] Inside particles, the scalars combine into a common information unit shuffled around in a series of microstates defining spaces in a wave pattern—Heisenberg uncertainty of wave-particle duality.

Waves

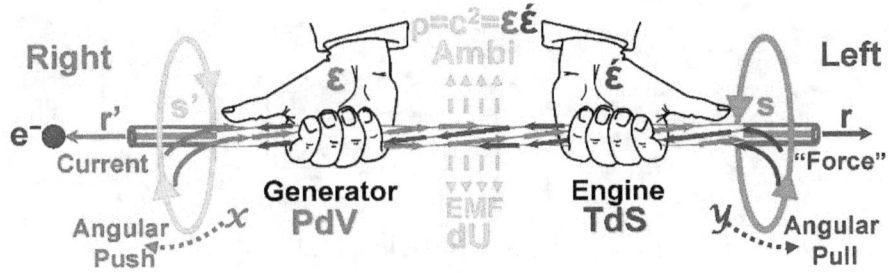

5.1: *Fleming's Rules & Force Variables*

Fleming's right and left hand rules show force directions in a complete process.[2] Linear forces are r' and r; angular are x and y; spins are s' and

---

[1] Jones, G.T. (2002). The uncertainty principle, virtual particles and real forces. hst-archive.web.cern.ch/archiv/HST2005/bubble_chambers/BCwebsite/articles/06.pdf.

[2] Daware, K. (2014). Fleming's Left Hand Rule And Right Hand Rule. electricaleasy.com/2014/03/flemings-left-and-right-hand-rule.html.

Quantum Relativity

s. The spins always exist with their angular and linear values losing their identities (order) to roles as radiant (EMR) and absorption (EMA) disorder.

Linear, spin, angular, right and left are specified for exactly the same reasons we specify A, B, and C in a quadratic evaluation: to identify which scalar applies to what. This is information reflected in B as the spinor equivalent. In the example, A is a unit of order (x). As an explicit intrinsic value (EMR) we can simplify to A=1. Available space for distribution is the EMA disorder function of y substituted in at C. This change shows why we can't over simplify to ABC. We must also accommodate abc≡A'B'C'.

$$Ax^2 + Bx + C = 0 \qquad x = \frac{-B \pm \sqrt{B^2 - 4AC}}{2A}$$

$$x^2 \pm 2yx + y^2 - c = 0 \qquad x = \begin{cases} \dfrac{\mp 2y \pm \sqrt{4y^2 - 4(y^2 - c)}}{2} \\ \dfrac{\mp 2y \pm 2\sqrt{y^2 - y^2 + c}}{2} \\ \boxed{\mp y \pm \sqrt{c}} \end{cases}$$

5.2: Scalars in Quadratic Equation Example

Classical force uses Newton units=kg m/s².[3] Quantum forces are presumptive scalars that can be converted into energy, momentum, and classical force easily (below). The universe utilizes attributed values as if they are real. Information-wise, they are like the 0s in binary. If a unit of energy is a complete information packet, then bits add up specifically, and all the 0s are just as significant as the 1s. Unlike bytes in binary, these information packets are infinitely divisible.

Classical: $F = ma = \left(\dfrac{mv}{\Delta t}\right)_{relativistic}$; $\Delta t = \left(\dfrac{r}{v}\right)^{orbit}$

QF of spin (light): $F_s^2 = \Delta F_L^2 + \Delta F_A^2 = \rho \left(\dfrac{\hbar}{c^2}\right)^2$, ρ=light surface

Frequency: $\upsilon = \left(\dfrac{F_s c}{h}\right)\left(\dfrac{1}{\sqrt{\hbar \varepsilon_v}}\right) \equiv \left(\dfrac{E = pc}{h}\right)$
cycles / second — attributed

Momentum: $p_x = \left(\dfrac{F_s}{\sqrt{\hbar \varepsilon_v}}\right) = \left(\dfrac{h\upsilon}{c}\right) = \left(\dfrac{h}{\lambda}\right) = m_x v$
attributed — attributed

5.3: Classical Force, Spin, Frequency, Momentum

Force conveys mechanically from one intrinsic field to another, or as light.[4] QCD colors (rgb) correspond with wave categories: intrinsic/ordered

---

[3] (Oct. 1, 2014). SI Unit of Force. UK: National Physical Laboratory. npl.co.uk/reference/faqs/si-unit-of-force.

[4] Henderson, T. (2018). Categories of Waves. physicsclassroom.com/class/waves/Lesson-1/Categories-of-Waves

red (j), disordered blue (i), and quasi-ordered green (h). Dashed arrow shows propagation direction. Solid arrows show mechanical displacement. Force surface is the phase area (A).

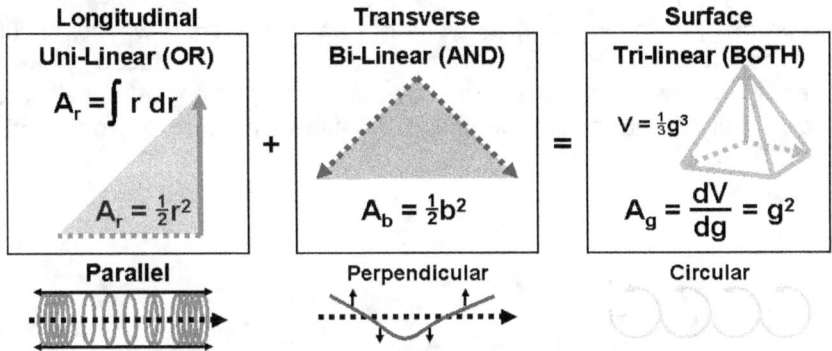

5.4: *Longitudinal, Transverse, Surface Waves*

Light only conveys as a transverse wave (disorder). This limits force associations with light to linear r and angular y. As in all wave information, one force acts as emission (EMR) while the other acts as absorption (EMA). When this information acts on matter, it attempts to normalize with the intrinsic information of that matter. The result is oscillation that can cause identity change, like flavor changing in neutrinos. [5]

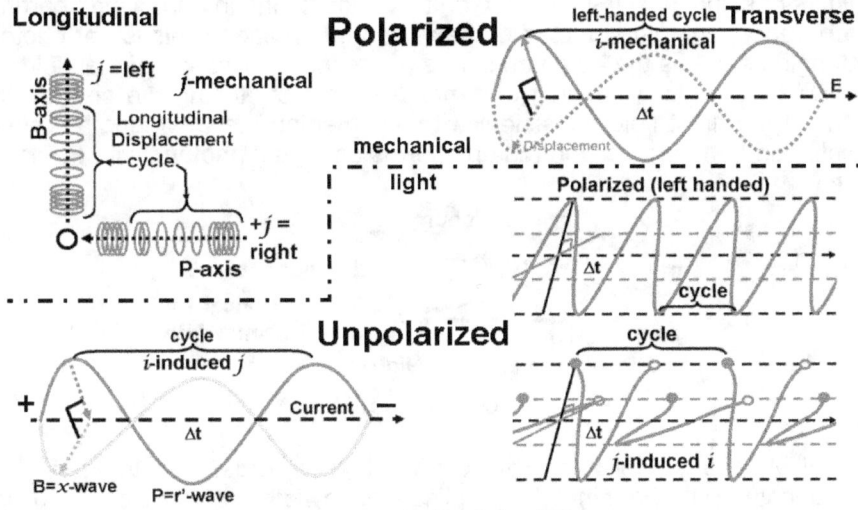

5.5: *Polarized/Unpolarized Waves*

Convention excludes longitudinal waves from polarization. Its polarization is chiral, with the B-axis distributive like sound. A gravitational

---

[5] Casper, D. (1998). Neutrino Oscillations. ps.uci.edu/~superk/oscillation.html.

wave would be longitudinal light (pure EMA) using spacetime density as its medium instead of matter. Haar "right" (e.g. gravity, e⁻ current) and "left" (e.g. heat or electrical force) are angular-B (cyan) and linear-P (red) relative axes.

The diagram provides three types of transverse (light) waves: circular, linear, and unpolarized.[6] "Circular" is $\hbar$-unpolarized (i induced j). A filter makes this "linear" = i-polarized (an unequal pairing of i =elliptical). Another filter depolarizes to j (induced i). These are commonly depicted:

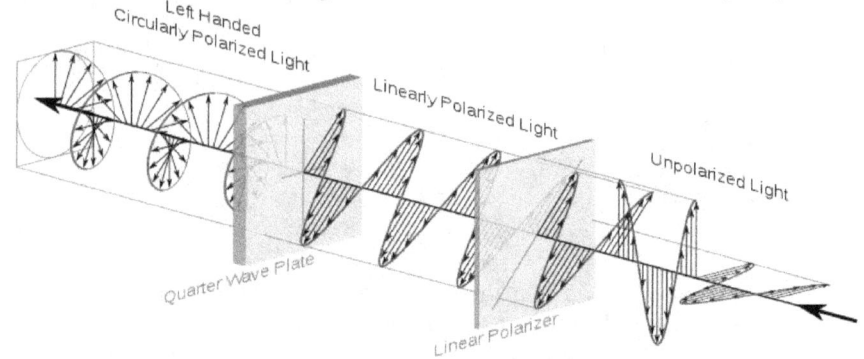

5.6: Filtering Polarization

Being without its own spacetime definition, light acts as a diverging field (signal/luminosity loss and drift[7]). Each point in that field, barring interference, contains the same shape and information. This is not photon entanglement. It is the common phase (moment) information of that field.

Divergence is a function of expanding void stretching the size of the points in the light field consistent with lengthening the wavelength. Those points can be interpreted in material terms as virtual photons with Planck's $E=h\upsilon$.

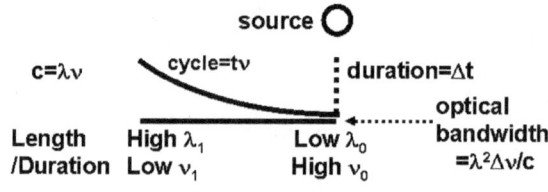

5.7: Optical Bandwidth

Initial energy followed by propagation divergence stretches the wavelength (λ) widening the optical bandwidth ($\Delta\lambda=\lambda\Delta\upsilon/c$).[8] Optical

---

[6] Erdogan, T. (2018). <u>Understanding Polarization</u>. Lake Forest, IL: Semrock, Inc. https://www.semrock.com/understanding-polarization.aspx.
[7] Rajaraman, R. (2010). <u>Antennas & Propagation</u>. NE University. ccs.neu.edu/home/rraj/Courses/6710/S10/Lectures/AntennasPropagation.pdf.
[8] Paschotta, R. (2007). <u>RP Photonics Encyclopedia: Bandwidth</u>. Germany: RP Photonics Consulting. https://www.rp-photonics.com/bandwidth.html.

bandwidth describes the surface of the wave passing through larger openings and around smaller.

To capture a low-energy distribution, a surface can be porous if the holes are smaller and general surface bigger than the optical bandwidth. The surface then reflects into focus or captures directly to observe.

Leaving an aperture open at length as to make a deep field observation increases the surface-time. The energy isn't accumulated all at once like a real photon, but instead accumulates to a virtual photon value.

These details are significant here because if light is to excite or otherwise modify matter, it has to be able to manifest in the available spaces of matter. Space availability depends on entanglements and available unused spaces. Weak bosons have very little available, where photons are conveyed at the same speed not easily allowing for accumulation.

This brings us to something waves (light) and bosons have in common: unit spin. The surface unit (radius) is cut in half when it evolves into or is applied to a volume. A volume more fully occupies a space. The creation of a volume tends to fill quantum numbers (e.g. change axes), limiting the ability to share that space (Pauli exclusion principle).[9]

Bosons are transitional (virtual) particles. They have a unit of spin because their spaces are not fully constructed, even though weak bosons liberally use volumes. Bosons can conditionally pass through the spaces of other particles.

The same is true with light. The unit cycle in frequency is the spin corresponding with a change function: unpolarized s' with j, polarized s with i, and circular ρ with h. Where things get awkward is in how much scalar value can occupy a space anyway. Of course that space has to be contained, so it isn't just a space, it is a spacetime.

Quantizing

Traditionally, the terms permeability and permittivity are applied only to EM. We are expanding them to fit their linear, angular, and spin analogs, plus the vector axis relation as a modifier class (subset V explained below). This modifier and EM have the unique quality of defining a fully occupied spacetime that contains change ($\acute{\epsilon}\epsilon = 1/c^2 = ij/\Delta$). The others define fully occupied spacetime in a change container (spacetime) $\acute{\epsilon}\epsilon = c^2 = \Delta/ij$.

Quantization is acquiring the proportional values required for a change unit identity. The values are evaluated relative to constants defining the proportional boundaries of spacetime. Spacetime is quantized by filling a permittivity (ϵ) or permeability ($\acute{\epsilon}$) equivalent quantum number. The constants (pg. 56) are specific to linear, angular, spin, and electromagnetism (distribution).

---

[9] Nave, C.R. (2017). Pauli Exclusion Principle. Georgia State University. http://hyperphysics.phy-astr.gsu.edu/hbase/pauli.html.

Quantum Relativity

The concept of permittivity generally describes the container into which value permeates (fills). Filling both together triggers Pauli exclusion to spontaneously discharge (e.g. photon emission). Roles do alternate with applications due to the evolutionary needs of matter. For example, degenerate density is numerically defined by the magnetic and linear permeability ($\mu_0/2G$, see pg. 152).

Exclusion applies when the value accumulates to $2Gc^2/\mu_0$. A lot can happen between ground state and this value, depending on how the degenerate's details are defined. A challenge of quantum forces is that they are in perpetual motion thanks to expanding void. Some degree will be lost, or have its information significantly restructured. The same degenerate can achieve S=0 (singularity) and switch to permittivity (Schwarzschild) as its baseline.

Even without quantizing, the constants are used in inverse square laws and other interactive applications. The forces as scalars can simply be given units by applying them to permeability and permittivity. Likewise, force in standard units (kg m/s²) can be conditionally reverse engineered to scalars or other unit applications. With Lie functions (see pg. 101) the ideal is either unitless scalars (f) or contextual kg m units $f_m = f\hbar/c$.

G & the Vector Axis Modifier

The vector axis modifier makes a rather substantive adjustment to Newton's constant. Both are confirmed by nuclear density=$U_A\mu_0/2Gk^2$ (pg. 152). There are a lot of issues with measuring G accurately from its long wavelength to interference of other fields including other forms of gravity.

G = 6.67384(80) E-11 m³/kg s² (CODATA, 2010)
    6.67408(31) E-11 m³/kg s² (CODATA, 2015)
    6.67545 E-11 m³/kg s² (Terry Quinn, BIPM, 2013)
    6.70197(08) E-11 m³/kg s²= $\hbar c^2\mu_0 \sqrt{2}/2\pi$ m² kg s

G
$i \Rightarrow 45°$
$r = \hbar c^2 \mu_0 / 2\pi$
[10]

5.8: Computing Newton's Constant (G)

---

[10] Mohr, P.J. et al. (2012). <u>CODATA recommended values of the fundamental physical constants: 2010</u>. physics.nist.gov/cuu/pdf/RevModPhys CODATA2010.pdf.
Newell, D. (2015). <u>CODATA Recommended Values of the Fundamental Physical Constants 2014</u>. codata.org/blog/2015/08/04/codata-recommended-values-of-the-fundamental-physical-constants-2014/.
Moskowitz, C. (Sep. 18, 2013). <u>Puzzling Measurement of "Big G" Gravitational Constant Ignites Debate</u>. scientificamerican.com/article/puzzling-measurement-of-big-g-gravitational-constant-ignites-debate-slide-show/.

$\varepsilon_V = 2\pi G\varepsilon_0/\hbar = 4\pi G/\hbar\mu_0 c = 2$ sec φ/m^2 kg s is the modifier as a proportion of the other permittivities, where φ is the relative rotation of the axes defined by the slope of the displacement curve $m_S = \tan \varphi$ ($=\Omega_\Lambda$ local cosmological constant density:critical density). The angular and spin axes become a spherical spacetime volume in change ($\hbar\mu_0 \rightarrow$ ms) $c^2 \rightarrow m^3/s$. This is converted to rectilinear form ($/4\pi$).

The rectilinear form is still in temporal phase. Unless you are applying a brane as volume-independent surface with temporal tension (Quantum Gravity type 1=QG1), you want the out of phase form. Out of phase is the circular radius, which is double the volume radius (2r).

To make G real, we have to rotate it to a right angle relative to the angular and spin axes. Rotation ideally converts from rectangular edge center to vertice (e.g. z=x+iy) at sec $\pi/4 = \sqrt{2}$. This would be an idealization of geodesics like a singularity. This radian is the rate by which space is curved into gravity, which is not consistent. The modifier includes both rotation and radial conversion (spherical 2r=circular R).

Quantization happens in different types of confining conditions. By confining we mean it cannot escape and ends up being focused and modulated into a material identity. The most common example is the creation of photons by electrons, or x-rays by protons. Weyl fermions, aka semi-metals are an example of this kind of ad hoc instantiation in a conventional alloy.[11]

We call such a perturbation **instantiation**—basically declaring matter where the energy conditions are focused and sharing a common information pattern. Sharing the common pattern is important and connects us to our next body of concepts: oscillation, macrostates, and microstates. Before we jump to that, we need to first recognize a quirk of matter and the universe.

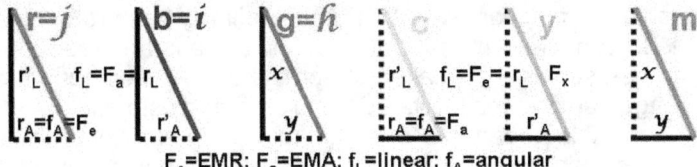

F_e=EMR; F_a=EMA; f_L=linear; f_A=angular

5.9: Triangulating Force in Color Distribution

The universe uses things it doesn't have. It also conserves them in such a way we can reasonably argue they are real. Annoying quantum existentialism aside, each quantum identity (color) has two fields: one taking the radiant value and the other absorption value. These triangulate into a common force value as if the universe wasn't already deceptive enough.

These identities are created simultaneously in chiral pairs or at least opposite orientations. This is done by accumulating scalar value in an

---

[11] Kuroda, K. et al. (Aug. 21, 2017). Evidence for magnetic Weyl fermions in a correlated metal. https://www.nature.com/articles/nmat4987.

available pair of related but opposite bands. Band nature determines whether the offspring will eject independently (with no effect on momentum as the ejection has no momentum), or split the host into two new particles.

This matter-antimatter pairing has created a degree of confusion. A photon, for example, interacts with its reflection, but is not created with its antiparticle. In other instances, chirals (imperfect mirrors) are created. Chirals work together to form ordinary matter. The differentiation occurs where chirality is renormalized (made relativistic) into helicity. There is no antimatter proportion violation, only misunderstanding.

Ordinary matter also has an antimatter form, but it isn't created the same way that requires both be created together. They also become individuals, and like any set of twins, have their own distinct lives, evolutions, and deaths. Evolution typically means blowing off enough disorder to be sustainably ordered. This is the simplest way to explain why the universe of conventional matter is mostly electrons and protons.

Back to our more primitive creations. Only a single unit of radiant value needs to accumulate to its proportional quantum number in this process. The proportion depends on the spacetime available for accumulation. When the quantum number is achieved, the change function of that space divides.

The division unzips the EMR-EMA values into distinct bodies of information. The sequence is influenced by the parent. The result is fraternal twins of opposite characteristics imperfectly mirroring each other. The ground state value of each identity is then a function of its EMR and EMA together.

This basic path of creation results in two parts defined in whole or in part by entanglements into one or more particles. Each of these relationships is defined by information specifying exactly how much value to apply to which field at which point in the sequence.

Each change function, whether it is defining a part or interaction, constitutes a container reacting to this information in sequence. The containers themselves are called macrostates. A distribution of micro energy changes among macrostates is called a microstate.[12]

When energy is added into this quagmire of interactions, the transient and native information are subject to normalizing (equilibrium 2nd Law of Thermodynamics[13]). The conflicting instructions result in oscillation that risks identity changes like neutrino flavor change. On a productive note, energy acting on the system changes momentum.

Even though microstates, macrostates, and oscillation were not designed for these quantum purposes, their concepts generally apply. Before we review, let us identify the key differences.

---

[12] McGovern, J. (2004). Microstates and Macrostates. University of Manchester.
theory.physics.manchester.ac.uk/~judith/stat_therm/node55.html.
[13] Redner, S. (2006). Equilibrium and the Second Law of Thermodynamics.
http://physics.bu.edu/~redner/211-sp06/class-macro-micro/2nd-law.html.

- QR macrostates are contextually proportional and specified by change containers (as opposed to arbitrary and equal).
- At least half the macrostates must have value in any microstate (as opposed to any one can contain all value).
- The lowest maximum value for a macrostate is 2 (maximum 6).
- True quantum microstates are a direct function of the information sequence. For QM simplicity (Bell's theorem), we generalize as if the information was perfectly sequenced (S=0).

Bose-Einstein

Boltzmann and Gibbs used an accessible density probability on the macrostates variable (ρ ln ρ) to conceptualize S=k ln W. Shannon used a binary information approach: $\sum p_n \log_2 p_n$.[14] Bose and Einstein took a simpler unit approach.

$$\omega(E_x, N) = (E_x + N - 1)! / E_x! (N-1)!$$

The Bose-Einstein microstate divides integer units of energy (E_x) among N equal but distinguishable particles. It assumes all the energy of the system can be contained by any single part, or distributed among them severally. Consider 6 distinct particles (N) with 9 units of energy (E_x). To compute microstates:

$$\omega = (9+6-1)!/9!(6-1)! = 14!/9!5!$$
$$= 14 \times 13 \times \underline{12} \times \underline{10}/\underline{5} \times \underline{4} \times \underline{3} \times \underline{2} \times 1 = 14 \times 13 = 2002$$

To show this distribution, simplify the microstates into macrostates. Each macrostate represents an indistinguishable distribution of energy units across the parts. In the diagram below, each dot represents one of six material identities. These are "loose" identities, meaning their interaction is not relevant.

Where the dot is placed vertically describes the amount of energy units applied to that indistinguishable part. Each box represents a macrostate—a traditionally arbitrary region. Within a macrostate of the table below, the sum total of energy in the system is always $E_x = 9$.[15]

In the diagram below, at the bottom of the box are all the identities without value. These are highlighted to draw attention to them. If we label

---

[14] Bain, J. (2014). Boltzmann Entropy, Gibbs, Entropy, Shannon Information. faculty.poly.edu/~jbain/physinfocomp/lectures/03.BoltzGibbsShannon.pdf.
[15] Nave, C.R. (2017). The distribution of 9 units of energy among 6 identical particles (et seq). hyperphysics.phy-astr.gsu.edu/hbase/quantum/disbol.html#c1.
Hock, K.M. (2014). Basic Statistical Mechanics. University of Liverpool. http://hep.ph.liv.ac.uk/~hock/Teaching/2013-2014/1-basics.pdf.

Quantum Relativity

each of these dots to make them distinct, we may label them parts A,B,C,D,E,F respectively. At the top of each box is a number indicating the number of possible ways to label the energy values provided.

All those with zero value aren't counted as ordered. Only those with value can be ordered. If all the energy is in one identity container, that container could be any one of the six identified, so there are six possible microstates for a macrostate (Ω) distributing all the energy to one. There are 30 ways we can arrange the containers when only two have energy, and so on.

Distinct Microstates (2002) per **Distribution** (26) of **Energy** (9) in Particles (6)

5.10: Bose-Einstein Microstate Example

N is the number of possible containers. N_O is the number of containers whose value is 0 (noting 0!=1). N_A is the number of containers with energy.

$$\omega/\Omega = N!/N_O!N_A!$$

Adaptations

These microstate diagrams are not to be taken as literal constructs. The parts are interacting to create spaces. We are simply trying to illustrate which parts are doing what when. "Structural" evolutions are covered under Quantum Morphology (pg. 105 et seq.).

QR analysis of microstates is about the use of space to interact, evolve, and move. Naturally this should be done accounting for information. This is generalized for QM convenience to S=0. For microstates to reflect matter in this process and the use of space, the distribution needs to reflect proportional values and maintain enough distribution for identity. We assume at least half the parts and interactions will always have some value and up to half can have no value.

How the parts apply those values when they have them is proportional to them specifically. For example, the ratio of $v:\mu$ is $1:e$. Since we are working in whole increments we round up to $1:3$. We double this to be sure we have something to actually distribute, but still a bare minimum.

$$S = -\underbrace{\left[k_B \ln \omega\right]}_{\substack{\text{unit conversion} \\ \text{microstates}}} \underbrace{\left[\rho_B = \sum_i \overbrace{\left(\frac{E_{xi}}{E_x k_i}\right)}^{\text{units applied}} \overbrace{\left(\frac{\omega_N}{\omega}\right)}^{\text{instances}}\right]}_{\substack{\text{total units} \mid \text{microstates} \\ \text{specific change} \mid \text{container weight}}}$$

$$\underbrace{}_{\text{APPLIED RATE}} \quad \underbrace{}_{\text{ENERGY DENSITY}}$$

5.11: S-entropy as Energy Density

We always reduce to the bare minimum. An identical or even chiral pair of anything like a photon can have distributions looking like 2,0,2, 2,1,1, 2,2,0, 3,1,0 etc. but annihilates at 4,0,0. Of course this means we really need to know the "structures" we are working with. Structure better fits Bose and Einstein's relativistic particles than the quantum particles they are constructed from.

Like Parts

			0 / 3		1 / 2		2 / 2 / 0
Photons	$\omega=10$ $E_x=4$ $S_E \approx 1+3$	4	013,103 310,301	3	112,121, 211	3	202,220, 022
Leptons	$V_E \approx 2.4$ $S_E \approx 1.6$	4	013,031 310,130	3	112,121, 211	3	202,220, 022

Unlike Parts

			0 / 1 / 3		1 / 2		0 / 2
Most Gluons	$\omega=5$ $E_x=4$ $S_E \approx 1\frac{1}{5}+2\frac{4}{5}$	2	103,013	2	121,112	1	022
Weyl fermion (type I)	$\omega=6$ $E_x=4$ $V_E \approx 2+2$	3	130,031, 0 3	2	1 2,121	1	022
Weyl fermion (type II)	$\omega=6$ $E_x=7$ $V_E \approx 4+3$	2	$1V_3 2$, $2V_3 1$	2	$2V_2 1$, $1V_2 2$	2	3 / 0, 0 / 3

5.12: Microstates of Like/Unlike Binary Virtual Particles

Analyzing microstates like this shows why particles move around as they do (see below). There is of course more to the analysis than this. Photons have the ideal ground state speed of c because they are just interacting with their reflections. They don't need momentum. If you add momentum, they suddenly have mass and slow down rapidly.

Oddities like singlets and weak bosons become incredibly messy. The three photon singlet (rc+by-gm)/√3 requires a unit in each rc, by, gm, up to 2 in rc and by, and up to 3 in gm. The distribution among primordials and bands comes to 558 microstates for E=5 following those rules.

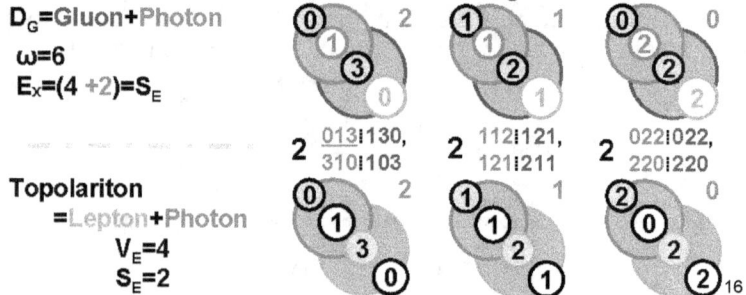

5.13: *Microstates of Singlet & Topolariton Virtual Particles*

Neutrinos consist of chiral halves synchronously entangled with isospin (antineutrino −½ reverses microstates causing annihilation with a +½ neutrino). This is exactly opposite to photon perfect. They have no intrinsic motion because their chiral halves undo each other. They oscillate and do get kicked around easily. Other leptons have intrinsic momentum, which is problematic for identifying their ground state values.

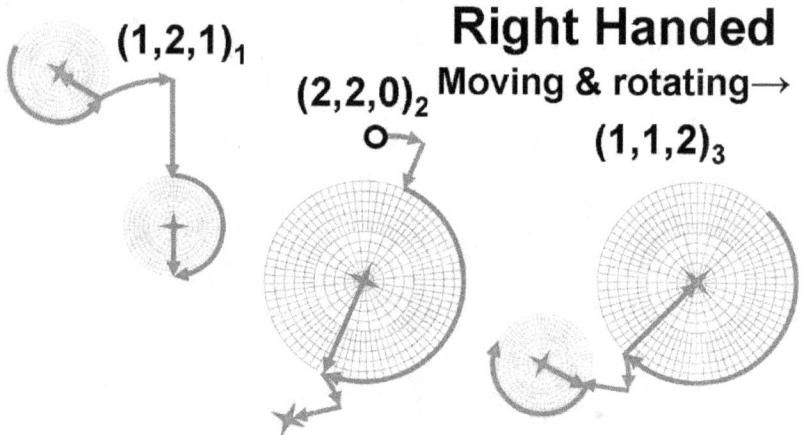

5.14: *Affect of Microstates on Intrinsic Motion*

The diagram above neglects the 3,1,0 position, but gives you an idea. It is also flat when the changes are three-dimensional. It is far more

---

[16] For more on topolaritons: Refael, G. et al. (Jul. 1, 2015). Topological Polaritons. CalTech. journals.aps.org/prx/abstract/10.1103/PhysRevX.5.031001.

ambiguous if you put in actual information instead of perfect sequencing. It is right handed because the space is defined in the same direction as the trajectory of intrinsic motion. As energy is moving around, it is adding value to the parts and interactions either together (i=AND or h=BOTH) or sequentially (j=OR) as locally defined.

This is a highly straight-forward system of simply increasing line and arc lengths per change function specifications. The energy is literally just moving around in a fixed pattern. The totality of that pattern sequence is a cycle, which only happens ONCE and then the universe happens to it until it is violated.

The cycle details are the individual. They are semi-independent of time in that the light is still being conveyed at the speed of light through all these microstates. As you add energy to the system, the resistance to completing microstates diminishes as time differentiates into entropies. This causes frequency (cycle rate) to increase (dilate) despite increase in band length.

The added energy comes with its own oscillating sense of direction. It can ONLY accumulate in these available spaces. The fewer available spaces, the harder it is to move something. Being difficult to hit the target is not the same as resistance of the target to change (e.g. mass).

When energy accumulates in these spaces, its sense of direction affects the trajectory of the microstate motion. For a neutrino that original direction is nowhere. Real easy to move: just add energy. The argumentative effect on microstates is appropriately called oscillation. Oscillation energy levels can cause identity change via proportions or new matter formation, as with an electron creating a photon.

Intrinsic Mass

These microstate conditions are incredibly ambiguous as we will see with particle morphology (see pg. 113 et seq.). As with our previous microstate analysis, this is superficial modeling. The real thing follows the information. These are hypothetical models for QM simplification. More importantly, they show where geodesics and mass emerge.

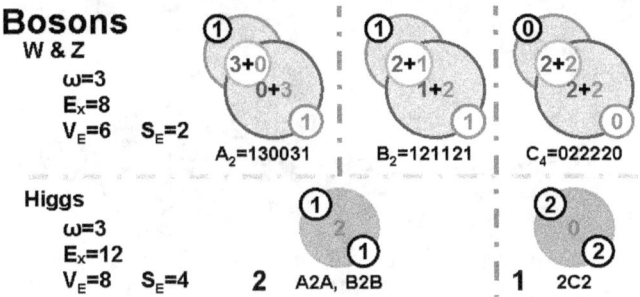

5.15: *Possible Microstates of Weak Bosons*

Quantum Relativity

Just as important as motion is the geodesic field equation (GFE) giving ground state mass. Relation to Newton and mass via Poisson is provided on pg. 20. The GFE requires volume and surface interaction. Values must simultaneously define those interacting spaces.

The frequency and degree these interacting as geodesics is a quantum mechanical way to define generic mass. This is obvious comparing weak bosons of consistent equivalent distributions with irregular and uneven distributions of charged leptons. The unreliable distribution robs them of most of their mass and leaves a lot of unused space. Weak bosons simply alternate which half of their parts is acting as surface versus volume at any given time.

The W & Z bosons shown here are partially covering up their other parts. The Higgs boson is easier because it is simply two chiral type II Weyl fermions in transition most probably to tau neutrino or related.

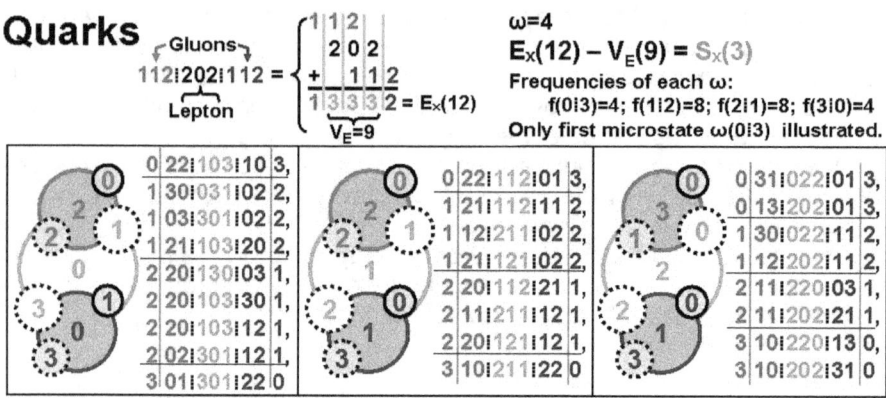

5.16: Possible Microstates of Quarks

Quarks are significantly involved because they consist of a lepton plus two gluons. This makes them highly volatile and unstable. When they form a baryon such as pion, kauon, nucleon, etc., they actually "bond" on several levels. The easiest way to visualize them is in layers, as with degeneracy layers of a nucleon.

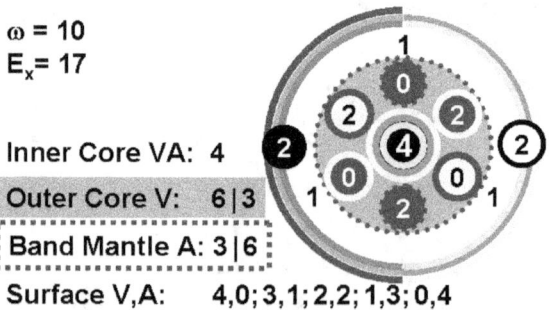

5.17: Possible Microstates of Nucleons

The gluons normalize into commensurate leptons. The arrangement of these leptons is even more interesting. A nucleon is seemingly held together by two tau neutrino leptons acting as both nucleating core and surface. It is as if the core is a Higgs except for significant and confined role changes.

Misapplication of the Higgs boson draws us to outline the elements of classical mass we are systematically resolving:
1. Volume and surface interaction satisfies geodesic field equation.
2. Microstate conditions modify quantum mechanically the degree and frequency GFE conditions are simultaneously satisfied.
3. Excess energy in available spaces becomes momentum adding mass to any identity via Relativistic momentum: $E^2=(mc^2)^2+(pc)^2$.
4. Temporal effect is defined in degrees by band rotation changes (next chapter), which directly affects the ability to satisfy GFE and mass.
5. Dilation and oscillation apply as bands approach their unit capacity.

Understanding QM interactions affecting GFE constructs will provide the solution to the "mass-gap problem" (Yang-Mills[17]). Mass and its related resistance to change in mass (temporal dilation) emerge together from the GFE. Resistance to mass change is resistance to changes in the GFE, which includes energy affecting momentum. The GFE apply when spacetime constructs from charge interactions enable a degree of non-neutralized surface and volume interaction.

We observe a third emergence of time as an interactive effect here. As an effect it resists: microstate cycles (frequency), GFE/mass change (dilation), and order-disorder (cosmic flux/thermodynamic arrow). Increase of energy in a system is decrease of free energy, increasing the temporal effect.[18]

The effect rate contracts the use of space (dilates) because time defines the linear increment of spacetime. Increased frequency affects how light and material microstates interact with their environmental spaces. Increased mass dilation reduces space available for energy to act on, decreasing efficiency to accelerate.

Like mass, time is real emerging form hypercomplex in degrees. It does not have a negative state. The arrow doesn't reverse. Some have misconstrued forming order as reversing time.[19] Reversing the effective rate of time is like draining water from a glass or bouncing a system back to ground state. The laws of Thermodynamics specify how that is done (focusing disorder in one place and order in another, see 2nd and 3rd Laws pg. 43 et seq.). Time is still resisting that order-disorder change.

---

[17] Douglas, M.R. (April 2004). Report on the Status of the Yang-Mills Millenium Prize Problem. www.claymath.org/sites/default/files/ym2.pdf.
[18] Tuisku, P., Pernu, T.K., & Annila, A. (Jan. 6, 2009). In the light of time. rspa.royalsocietypublishing.org/content/465/2104/1173.
[19] Eck, A. (Nov. 28. 2017). Scientists Reverse Arrow of Time in Quantum Experiment. pbs.org/wgbh/nova/next/physics/scientists-reverse-arrow-of-time-in-quantum-experiment/.

6. Constructing Spacetime

To see how microstates, momentum, and mass work, we first need to understand how the use of space evolves to give us workable fields. Relativistic spacetime emerges when the GFE apply in a complex multi-stage developmental process.
1. Extra-temporal color charge surfaces are valued.
2. Temporal phase impositions are applied to define fields with potential to strongly interact.
3. This is confined in a quasi-temporal interaction such as
   - bond creating a subspace volume (flavor)
   - entanglement creating a subspace surface
   - transient (weak) interaction creating both surface and volume from the same interaction satisfying GFE.
4. Transition resolves, volumes and surfaces confining together into a common attributed entanglement surface conditionally satisfying the GFE.

Background

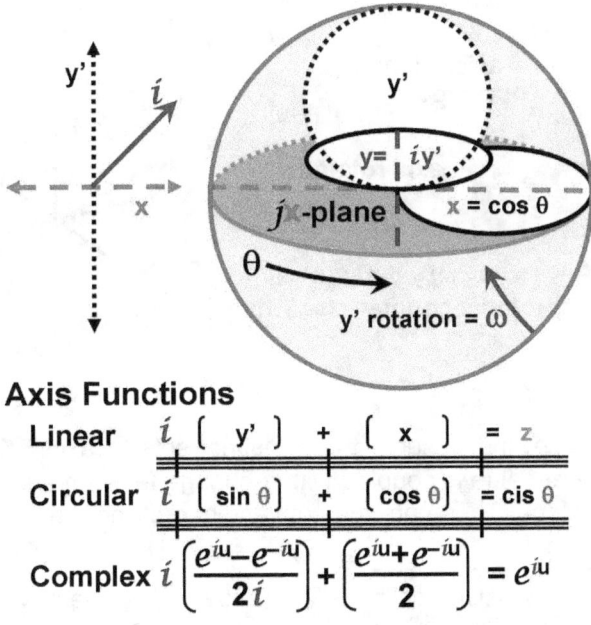

6.1: Complex Functions & Axes in Sphere

Quantum Relativity

Euler's $z=x+iy$ interprets in a variety of ways depending the context you put it in. His famous interpretation was of a helix due to application in rectangular (Euclidean/Cartesian) coordinates (see pg. 70). Euler's helix works for evolving multiple dimensions due to the conversion conventions sin u=y and cos u=x that enable making them a circular problem.[1]

There is only one small problem with the way these variables are discussed. The y' axis is imaginary due to the definition of sine. This of course applies to a 2-D coordinate system, which is ideational and therefore virtual/imaginary anyway. To show this, we use y' above to rotate into real y.

By multiplying iy', the imaginary element is stripped from the denominator of Euler's definition making it quasi-real. The only thing left imaginary in Euler's functions thereafter is the exponential use of i. Assuming u is a scalar, $iu=f(jx)=f(\theta)$ shows i acts here as an adaptive radian providing real results. Sine and cosine in polar coordinates provide circles straddling $\pi/2$ and 0 respectively as illustrated.

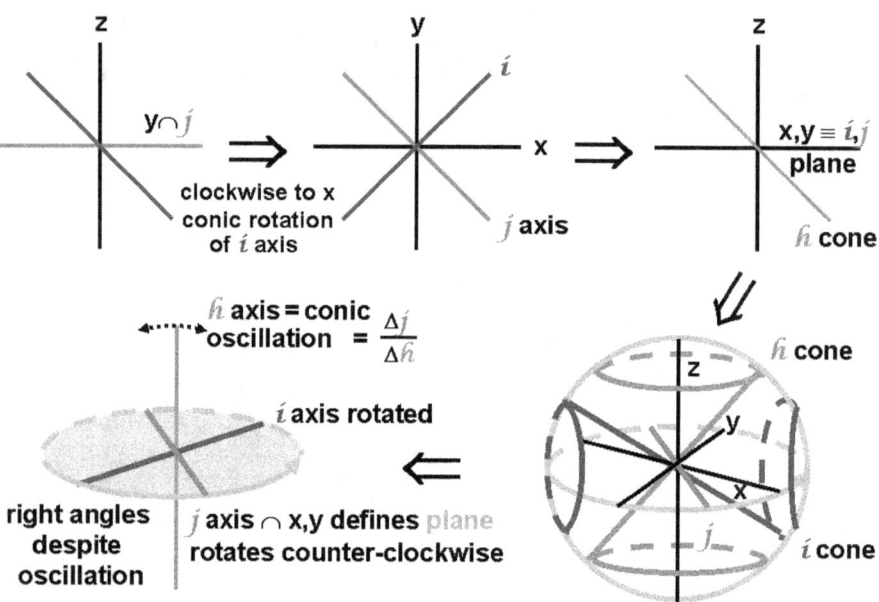

6.2: *Axial Rotations of Complex Change Variables*

Let us pause and observe three distinct sets of axes: $j \cap x$, $i \cap y$, and $h \cap z$. Each linear value is coupled with a change function such that an axis never occurs alone. These are the convenient rectilinear axes we are used

---

[1] Brown, J.W. & Churchill, R.V. (2009). Complex Variables and Applications. 7 ed. McGraw Hill. math.unice.fr/~nivoche/pdf/Brown-Churchill-Complex%20Variables%20and%20Application%208th%20edition.pdf.

to. They are not how nature does things, but thanks to the complex variables, easily made compatible. The first step to making them compatible is to recognize their 4-dimensional form.

If we follow through with applying our imaginary numbers across the board with the correct variables, we get an imaginary circle pretending to be real.

$$(ℏr=ȷx+ıy)^2 = (-r^2=-x^2-y^2) = (r^2=x^2+y^2)$$

Next, we combine these to form a REAL volume, including a z-axis. Correctly applied, we get a real sphere:

$$(ℏr+ℏ'z=ȷx+ıy)^2 = (z^2-r^2=-x^2-y^2) = (r^2=x^2+y^2+z^2)$$

Spherical $ℏr$ is a real point density distribution (divergence) of a generalization (gradient) in time (dt= $δȷ + δı$, see pg. 21). It derives from a hypercomplex n-Euclidean dimensional function that derives from a complex surface. It satisfies the elements of the Laplacian operator,[2] providing a common frame of reference.

In this system, z is an elevation relative to the surface xy on r. We now have our initial conversion conditions. The next matter of conversion is adapting the axes of the primary manifolds into this system. Naturally we have to start with similar axis basics recognizing XZ pairs with $ȷ$, and WY with $ı$.

We start with the assumption each root is a linear axis at a right angle to the others. We can then use the Haar directional system to create two pairs that ultimately set all four at right angles to each other.

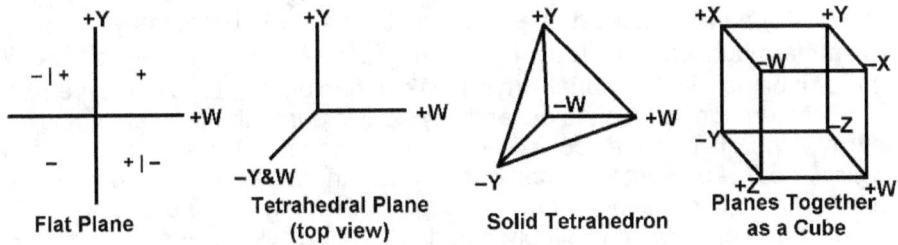

6.3: *Haar Linear Axis Relationships to Cube (4-D)*

Haar orientations are vital to Lie theory and manifolds (see pg. 96 et seq.), as they form a locally compact group as opposed to subtractive – and additive +.[3] Here we take a simple conventional rectangular plane and convert it first to Haar. This means positive and negative of the same axis are at right angles relative to each other.

---

[2] Stewart, J. (2017). Calculus: Concepts and Contexts – Enhanced. 3 ed. Phoenix, AZ: Content Technologies, Inc.
[3] Abbaspour, H. & Moskowitz, M.A. (2007). Basic Lie Theory. World Scientific.

Quantum Relativity

The general axes are also at right angles to each other. This is called a tetrahedral plane because we can connect all these points into a tetrahedron. We can also use an arc to connect them singularly to show an S-shaped pattern like i-entropy—e.g. electron orbit pattern.

Putting two planes of two Haar measure axes at right angles to each other connects four spatial dimension axes to the vertices of a cube. While impractical by human standards, this system allows a fluid relationship among related axes and a way to put them meaningfully together without conflicts. When nature fills a space it starts at a logical place. This system provides those places. They can fill sequentially or simultaneously even without shaping them.

Shape is where nature detours dramatically from our convenience. The first step in shape is the two-dimensional surface version of each manifold axis. For j and i, these are the planes illustrated earlier, which then apply to a sphere in phase.

Hypercomplex h is a circle whose center is not equal to the origin point of reference. This adds a layer of ambiguity. Instead of coming into form as a sphere, it first rotates into a toroid whose surface translates into that of the in-phase sphere.

Surface is the default assumption of each manifold, but the shape is undefined. Each surface draws out from its relative axis. It then rotates into its next shape. Rotating into shape is rather misleading as one really needs a volume as reference for that surface to shape to. That volume could simply be the force of a entanglement trapping into a common axis as we did with the radius and xyz axes above.

Volume derives by proportionally intersecting the axes of the change operators. The result is a strong bond creating a Weyl-fermion/lepton flavor. At a primordial level, the only way to join $j+i$ to get h is by stepping down the magnitude of i to its light form. To h, j and i are viewed as unit values it binds with individually (type I Weyl fermions) or together (type II).

The quantum rotation is deceptively equivalent to the term spin. It is an extra-temporal direction, so it is not like the rotation of a planet. It is a cycle, and it represents the generic value of the object as a whole. It is a quantum number relating to how the space is actually being used.

At a unit, the space is simply a surface. When it is a volume, the radius and unit spin are cut in half. Defining a volume is a quantum number unto itself, and a bit tricky as we will learn later. For now, the half-spin resulting from a volume flavor creates a Pauli exclusion. This space cannot be occupied by another volume.

There is one very reasonable complaint of mathematics, quantum theory of this level, and computers. Every tiny detail counts. Exclusion of other volumes does not mean exclusion of other surfaces or of volumes that are somehow chiral. A chiral volume would be an imperfect mirror like the tetrahedron planes. Because they can be set at right angles to each other, they can conditionally appear to violate exclusion.

Basic Interactions

Interactions evolve with matter. For now we are just doing basic spatial reconstruction. Primordial interactions include: bond, entangle, and transient. The differences are whether the axes intersect at some point in phase, at the edge of out of phase, or at an intermediary point.

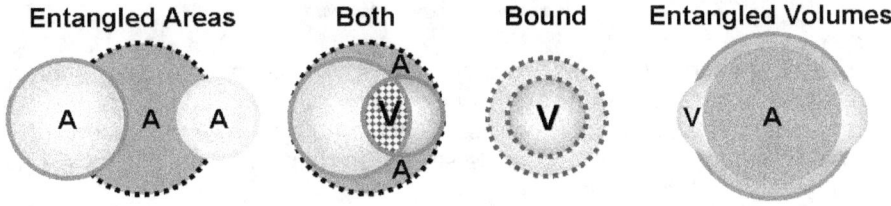

6.4: Interactions Compared

Bonds allow change functions to merge into a common self-defining volume and energy value. Without a bond, a primordial is a surface conditionally subject to phase condition applying it to volume without defining a volume. Primordial "bonding" is required to satisfy volume (flavor). Volume is also faked by the transient intermediary interaction.

Transient interactions are not bound or entangled but either and both. It is transient because it will resolve establishing bonds and entanglements. Faking volume does not trigger exclusion, but dramatically affects the GFE and mass. It can still incidentally pass through all or part of the same location of another object. The bond does not need to actually exist to be in a bond-like position and create volume.

Unlike bonds and entanglements, the transient interaction does not require energy reduction because all the parts are still individuals. It will certainly lose a lot of energy resolving its transience. Unlike full bonds, the transient volume is extra-temporal allowing for a sort of quantum tunneling—an effect of the Heisenberg uncertainty principle.[4]

Since the universe prefers the least restrictive path, to narrow the uncertainty, control the potential for a particular outcome. The mechanism for this is built into the extratemporal characteristics of matter like microstate cycles, and out of phase brane definitions. The universe cheats where it can to utilize these. In the case of quantum tunneling, it passes through a barrier it normally couldn't by not being excluded.

A fixed volume reduces the radius used to define the surface by half. In this case, parts of the volume and surface are swapping places in microstates. The surface does not exclude, so the surfaces permit passage. Consequently, weak bosons are given a unit of spin, which accounts for their surfaces. Their incredible masses result from the confined volume of transient interaction.

---

[4] Mastin, L. (2009). <u>Quantum Tunneling and the Uncertainty Principle</u>. http://www.physicsoftheuniverse.com/topics_quantum_uncertainty.html.

Quantum Relativity

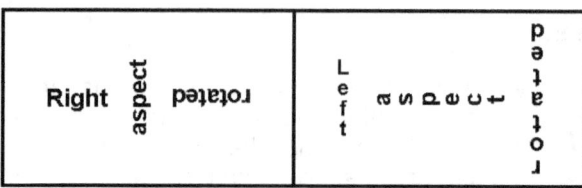

6.5: Rotating Aspects Example

This cheating includes rotating aspect position. As the diagram illustrates, aspect is the universe's way of saying which end is up, like our four axes intersecting cubic vertices. Without this universal recognition, space would act differently on fields simply by rotating them. The forms are already chiral, which helps.

Light cannot tell which end is what when it encounters a surface out of phase. It gets bent by the conversion of the space into phase, but otherwise ignores the surface. Light can only act on an available surface in phase. A singularity has an available surface, except it is put into phase relative to an enfolding surface, which conceals the availability to light.

Brane Surgery

QR branes are not to be confused with String branes. In QR a brane is a surface without depth used typically to generalize spatial definitions like the derivative of a volume or surface with depth (membrane). The space of a lepton is best described as a membrane because it has depth. Photons and gluons, however, consist exclusively of branes.

Branes are evaluated both in and out of phase—both contextually apply simultaneously. Axes can bond or otherwise define volume intersecting in phase. Entanglements are at the edge of out of phase definitions. The sequential difference in the branes of blue (i) and red (j) colors shows why they are left and right handed.

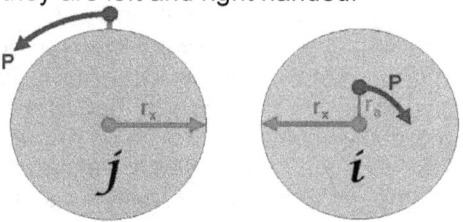

6.6: j-Right i-Left Branes

These are not, however, accurate representations of a brane surface being converted. The change function divides it into separate linear and angular values handled differently. For j, the rotating radius is stacked onto the linear. For i, both are measured from the geometric origin. The brane out of phase is ellipse-like (smoothing computations) rather than circular.

Quantum Relativity

$$1 = \frac{x^2}{a^2} + \frac{y^2}{b^2}$$

$$a^2b^2 = b^2x^2 + a^2y^2$$

$$a^2b^2 = b^2\cos^2 u + a^2\sin^2 u$$

$$y = \sqrt{b^2 - \frac{b^2x^2}{a^2}} = b + j\frac{bx}{a}$$

6.7: Elliptical Functions

The difference makes them as easy to handle as applying a projection like the WMAP image below to the surface of a sphere by wrapping it around the equator. A singularity is extremely convenient to exactly this evaluation method. The wrap into phase distinguishes linear poles from angular equator resulting in an imperfect sphere.

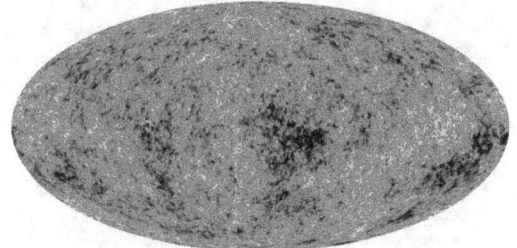

6.8: NASA/WMAP CMBR Elliptical Image (#121238)

Phase entropies (h and h') are also extra-temporal and subject to being put into temporal phase (k_t). Their branes are a tad more abstract in their construction. So far we have only dealt with linear and angular divisions of i and j. Put together we have linear on linear defining angular on angular. Thanks to j they are stacked, and thanks to i they are concurrent.

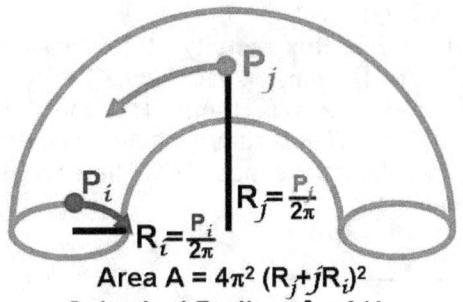

Area A = $4\pi^2 (R_j + jR_i)^2$
Spherical Radius $r^2 = A/4\pi$

6.9: Toroidal Functions

The values going into the toroid are angular. You have to backtrack from angular to linear radii to find the toroid's surface area. You can then derive the spherical radius that surface applies to.[5] Magenta already corresponds to longitudes. Rotation of green into temporal phase provides latitudes as shown.

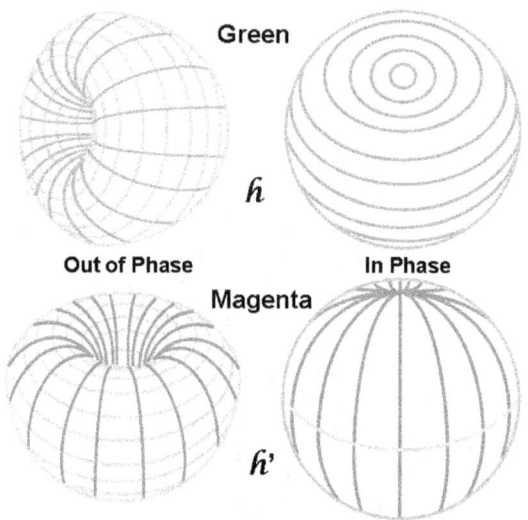

6.10: *Toroids Rotating into Phase-Spheres*

There is an extra layer of ambiguity in the out of phase mode relevant on two levels: strong interactions and magnetism. Strong interactions fill available space, like i into j. This is a reason green/magenta are always part of a strong primordial bond. The energy proportions are why the bonds must be simultaneous.

A magnetic field is toroidal like out of phase magenta (h'). This is misleading as it is both, but we like orienting it this way. It is toroid expanding latitudes by intrinsic ordered j and diverging longitudes by disordered i. Their common function is in units of angular pressure (A/m=2×10⁻⁷kg/m s²=$\acute{\epsilon}_A$/2).

In a magnetic field, B is **flux density**—the intrinsic energy value= $f(j)$. Its units are kg/As², with 1A (amp)=2×10⁻⁷ kg/s².[6] B/μ₀ makes it an angular pressure unit ($\acute{\epsilon}_A$) compatible with H=A/m. H is the **field strength**, better described as angular flux= $f(i)$. B/H=μ_m is material-specific permeability[7]—how energy is giving value to a space. The difference with angular is how the geometric origin is used.

---

[5] Ness, R. (2015). Surface and Volume Equations. nessengr.com/technical-data/surface-and-volume-equations/
[6] SI Brochure: The International System of Units (SI) [8th edition, 2006; updated in 2014]. bipm.org/en/publications/si-brochure/ampere.html.
[7] Nave, C.R. (2017). Magnetic Field Strength H. Georgia State University. http://hyperphysics.phy-astr.gsu.edu/hbase/magnetic/magfield.html.

Inactive Surfaces

An entanglement space is given value by the absorption values of the parts at ground state plus excess value in the system. It increases as the energy of momentum is added causing the space to stretch like a rubber band. When that energy is released, the "band" snaps back to its ground state value. The ground state value is necessary just for conveying intrinsic energy of identity through the microstate process defining a cycle.

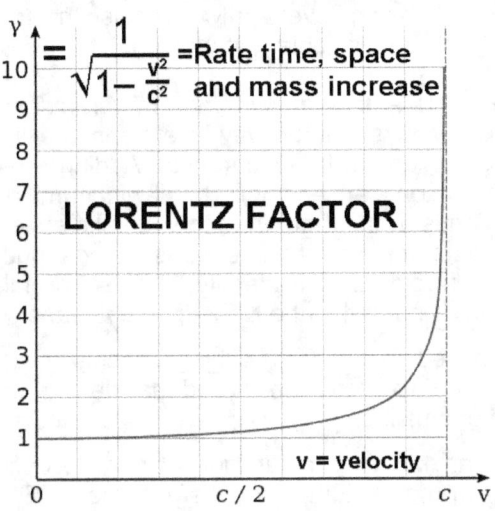

6.11: *Lorentz Dilation Function & Graph*

The energy of momentum (p) cannot exceed the available absorption value of the parts, despite the lack of apparent boundary conditions in $E^2=(mc^2)^2+(pc)^2$. As that energy accumulates, the microstates defining the intrinsic state of motion increase their frequency. That increase causes acceleration resisted by the inability to add more energy computed as dilation using the Lorentz factor.[8]

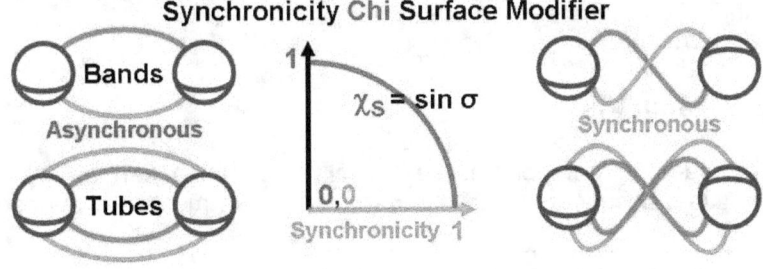

6.12: *Band/Flux Tube Synchronicity (Chi Twisting)*

---

[8] Forshaw, J. & Smith, G. (2014). <u>Dynamics and Relativity</u>. Hoboken, NJ: John Wiley & Sons.

Quantum Relativity

Bands always occur in chiral pairs of one change function—like i and i' in an electron. Bands are always of primordial magnitude. More advanced matter like the entangled volumes of quarks and leptons simply compound a pair of bands into each flux tube. Flux tubes thus have two surfaces in phase making them membrane-like with depth. This is a pseudo-geodesic affecting mass.

To be in temporal phase and act as a surface, the bands must be asynchronous—not cross over each other. This occurs with gluons and charged leptons. Crossing over only cancels the ability to function temporally (a synchronous chi=χ_S function). It does not cancel the entanglement.

Photons are colors entangled with their mirror image, where neutrinos are entangled chiral pairs. Either way, the parts are in opposite aspect positions causing the bands to synchronize. Adding energy or filtering to change relative rotation causes them to unwind and acquire relativistic qualities like dilation, slowing them from their ideal c-speed.[9]

Accelerating a neutrino adds mass to resist and triggers oscillation that can change flavor (Weyl fermion identity).[10] This was falsely attributed to ground state mass as quoted in the Nobel press release:

> The discovery led to the far-reaching conclusion that neutrinos, which for a long time were considered massless, must have some mass, however small.

A reach too far. Mass as a function of particle definition consists of parallel concepts of momentum and satisfying the GFE. To have mass at ground state, the GFE surface-volume interaction must be satisfied. Synchronous bands neutralize the surfaces making ground state mass impossible for neutrinos. The Nobel Prize was given to

> Takaaki Kajita in Japan and Arthur B. McDonald in Canada, for their key contributions to the experiments which demonstrated that neutrinos change identities. This metamorphosis requires that neutrinos have mass.

It only requires mass to oscillate into another flavor, not to exist. Beware false attribution errors and assumptions.

Tensor Manifolds

From a quantum perspective, absolutely nothing can be taken for granted. Just because you have an area does not mean it is shaped this

---

[9] Giovannini, D. et al. (Feb. 20, 2015). Spatially structured photons that travel in free space slower than the speed of light. Science. Vol. 347, Issue 6224, pp. 857-860. http://science.sciencemag.org/content/347/6224/857.

[10] The Royal Swedish Academy of Sciences. (Oct. 6, 2015). Press Release. nobelprize.org/nobel_prizes/physics/laureates/2015/press.html.

way or that or has a perimeter. Likewise, stating a volume does not equate to a geometry or surface. Each of these things is distinct. A distinct thing can be quantized—made a full quantum number with profound impact on the rest of the set.

A tensor is transformational scalar, vector, or group;[11] a geometric object of nth-rank (directions) generalizing scalar (n=0), vector (n=1), and matrix (array/atlas; n=2) transformations.[12] A manifold is a topologically Euclidean space—meaning it can differentiate as flat like Earth being round versus our perspective of a flat surface.[13] A tensor manifold here is a fundamental subspace construct describing a differentiable and interactively shaping field space.

Term Declarations

As stated in the introduction, to solve evolved problems we need to use evolved thinking, language, and concepts. We are here in this discussion due to a haunting mathematical experiment from 1991. The experiment sought a relationship between Euler's geometries, Relativity, and basic mathematical concepts like line, angle, scalar, etc.

The experiment found what has come to be called the Periodic Matrix or simply the Matrix. It was little more than a toy until several years later the same phenomenon was found with established elements in the social sciences. It has since evolved many times as its applications and terms were explored. To be safe, we mostly use it to organize variables.

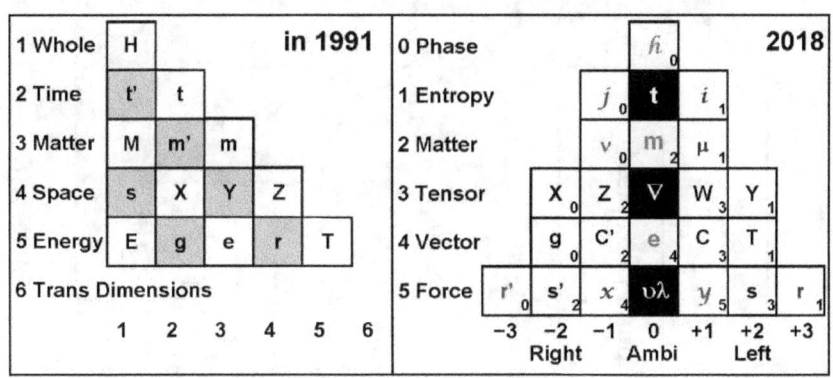

6.13: *Evolution of Variables Matrix*

---

[11] Kolecki, J.C. (2002). An Introduction to Tensors for Students of Physics and Engineering. Cleveland, OH: Glenn Research Center. grc.nasa.gov/www/k-12/Numbers/Math/documents/Tensors_TM2002211716.pdf.
[12] Rowland, T. & Weisstein, E.W. (2018). Tensor. From MathWorld--A Wolfram Web Resource. http://mathworld.wolfram.com/Tensor.html.
[13] Rowland, T. (2018). Manifold. MathWorld--A Wolfram Web Resource, created by E.W. Weisstein. http://mathworld.wolfram.com/Manifold.html.

Quantum Relativity

By established in the social sciences we don't mean it was just the theory of the day. It appeared in epistemology and developmental psychology concepts spanning recorded history. Geography and history were of no consequence. The same exact concepts emerged in exactly the same patterns independently everywhere.[14]

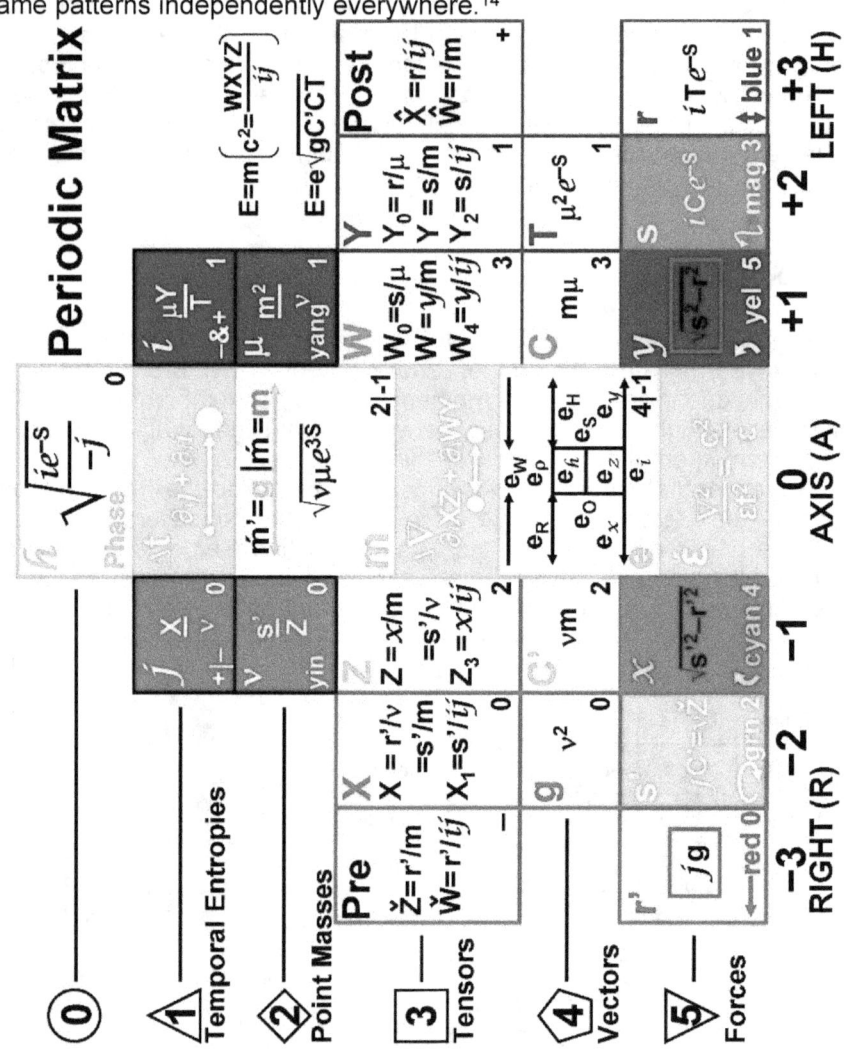

6.14: Periodic Matrix

Special note: We do not fully understand the Matrix and it is far from established in physics. We use the Matrix to keep track of all the variables and sub-variables and observe the mathematical behaviors of the variables interacting. When we use or create another graphic, we edit relative to this

---

[14] PüMa Tse. (2015). Love and Consciousness. Akademé.

Matrix. The Matrix is our most current and accurate. This is not always as easy to read as we would like. We are still working to solidify i (imaginary) is blue and i' (quasi-real i-prime) is yellow.

The initial relationship defining the Matrix simply followed Relativity using row and column values as exponential functions. In this way we can see the vector of gravity (g) derives from putting linear force r' in j-change: g=r'/j. We will revisit how we evaluate variable relationships on the Matrix at the end of the book (pg. 173 et seq.).

These are PDE variables, meaning they are unknown change functions.[15] They are very tricky and easy to misinterpret. Context changes which interpretation to apply. A seemingly straight-forward $ij\equiv t^2$ also evaluates as $i^2+j^2=2h^2$ with both Pythagorean and quadratic paths not to mention differential $\delta i+\delta j$=dt and the related extra-temporal phase interpretations. The interpretation depends on context, and generally all interpretations have some simultaneous role.

Sub/pre-applications are intrinsic buildups from a lesser level. At a fundamental level this would be a scalar (tensor of n=0). Super/post-applications are contributions to a greater level—another scalar. Interactions of these sub and super applications would result in mutual vector interaction (n=1).

One set of evaluations provides the Lie groups needed to construct and evaluate subspace manifolds, their interactions and evolution. We cannot presume, however, that each evolution should be broken down the same way. One great reason for caution is the mathematical expectation of smoothness, which evaporates quickly through the evolutions of matter.

Despite our handicaps, we find the processes and variables of the Matrix useful. To put meat on our madness, we observe manifolds on the Matrix cross over from one level of development into another. These are shown to the right and left as sub-Z|W (crowned as $\check{Z}$) and super-X|W (capped as $\hat{W}$). We also encountered it earlier with scale issues between interacting order and disorder requiring energy change in fusion.

Intrinsic spaces build into a complex atlas (array/matrix) of bi and n-directional interactive geometries.[16] Analyzing one force at a time as a Lie group enables us to construct these spaces and then apply them as interactive functions.

Before we step away from the Matrix as source we need to make a couple observations. The left handed skip in sequence contains the temporal boundary of $e^S=h^{-2/3}$. The top and bottom of the Matrix are fundamental. Everything else derives from their constructs starting with defining tensor manifolds. It is also unclear beyond the Matrix as everything ends up snapping back into this structure.

---

[15] Grigoryan, V. (2010). Partial Differential Equations. UC Santa Barbara. web.math. ucsb.edu/~grigoryan/124A.pdf.

[16] Welbourne, E. (Mar. 18, 2017). Atlases and Charts of Smooth Manifolds. http://www.chaos.org.uk/~eddy/math/smooth/atlas.html.

Quantum Relativity

Lie Groups

A Lie group is a differentiable manifold with smooth group operations, providing a continuous symmetry.[17] In a Lie bracket group [u,v], u is the function of change applied to v. Complex definitions like B±C form a chain, and a cross symmetry emerges as illustrated.

$$[A,B+C] = \mathcal{L}_A B + \mathcal{L}_A C = [A,B] + [A,C]$$
$$[A,B-C] = \mathcal{L}_A B + (\mathcal{L}_A C)^{-1} \quad\quad C = F/(v|\mu)$$
$$= [A,B] - [A,C] = [A,B] + [C,A]$$

6.15: Lie Algebra Basics

We first need our variables defined. Each primordial identity has two fields. Generically we can break these down using manifold variables. Each manifold is contained by change ($j\hbar i$) or identity ($v m \mu$) functions and given value by linear r'|r, angular x|y, or spin s'|s forces.

j-Tensors $\quad\quad h_j \quad h'_i \quad\quad$ i-Tensors
$f(r'|s'|x)=[V_f,\pm(m_f-v_f)] \quad\quad\quad f(r|s|y)=[\pm(V_f\pm\mu_f),m_f]$

$$\check{Z} = \frac{r'_M}{m} \quad\quad \text{Void Range} \quad\quad \frac{r_M}{m} = \hat{W}$$

$$\check{W} = \frac{r'_M}{ij} = V_{+L} \quad\quad \text{Brane Vector} \quad\quad V_{-L} = \frac{r_M}{ij} = \hat{X}_0$$

$$X = \frac{r'_M}{v} = \frac{s'_M}{m} = X \quad\quad \text{Brane Range*} \quad\quad Y = \frac{s_M}{m} \; ; \; \frac{r_M}{\mu} = Y_0$$

$$X_1 = \frac{s'_M}{ij} = V_{jS} \quad\quad \text{Spin Vector} \quad\quad V_{iS} = \frac{s_M}{ij} = Y_2$$

$$\frac{s'_M}{v} = Z = \frac{x_M}{m} \; \substack{\text{Volume} \\ \text{Domain*}} \; \frac{y_M}{m} = W \quad\quad \frac{s_M}{\mu} = W_0$$

$$f_M = \frac{cf_U}{\varepsilon_S} \Rightarrow \text{kg m} \quad Z_3 = \frac{x_M}{ij} = {}_{-A}V_{+A} = \frac{y_M}{ij} = W_4 \quad \text{Vector}$$

$$\text{Equatorial Range} \quad \frac{x_M}{v} \stackrel{B}{=} \hat{X}_4 \stackrel{H}{=} \frac{y_M}{\mu} \quad \text{Polar Domain}$$

*independent: Y_0= brane domain $\quad W_0$=range

6.16: Lie Elements of Sub-manifolds

---

[17] Warner, F.W. (1984). <u>Foundations of Differentiable Manifolds and Lie Groups</u>. Glenview, IL. wisdom.weizmann.ac.il/~dnovikov/Manifolds5775/arner_Foundations_of_Differentiable_Manifolds.pdf.
Abbaspour, H. & Moskowitz, M.A. (2007). <u>Basic Lie Theory</u>. World Scientific.
Miller, W. (1968). <u>Lie Theory and Special Functions</u>. New York: Academic Press. https://www.ima.umn.edu/~miller/lietheoryspecialfunctions.html.
Ebrahim, E. (May 19, 2010). <u>A Self-Contained Introduction to Lie Derivatives</u>. web.math.ucsb.edu/~ebrahim/liederivs_tame.pdf.

Quantum Relativity

For simplicity, the force value is already put into kg m units for each function. A significant number of these subspaces depend on emergent material values (m). These values do not occur without interaction we would consider confining into a greater identity.

In an ironic way, at a quantum level, relativistic is imaginary. This leaves range and domain values unspecified—unsustainably virtual perturbation. To fully utilize this system of mathematics will require us to explore how matter is assembled over the next chapters.

Now that we have our variables, we can batch them into Lie function groups by manifolds. Each material identity has two fields. Red=v and blue=μ chirals simply flip their negatives. The equivalent action for green flips manifold identities (X becomes Y, Z becomes W). Chirality only flips action sequence (e.g. u on v) for green. Left-handed change is always complex (u=A±B), and right handed is simple (u=A). The numbers under the manifolds are their sequence.

$$v \begin{cases} f(r') = [\check{W},\check{Z}-X] = [\check{W},\check{Z}]-[\check{W},X] \\ _{0\,g} \phantom{=[\check{W},\check{Z}-X]=} {}_{-1\,-2}{}_{-1\,0} \\ \text{RED } j\!\to\text{ v. Anti-RED Cyan } j'\!\downarrow \\ f(s') = [X_1, X-Z] = [X_1,X]-[X_1,Z] \\ _{2\,C'} {}_{+1\,0}{}_{+1\,+2} \\ \text{GREEN } \hbar\!\leftrightarrow \\ f(x) = [Z_3, Z-X_4] = [Z_3,Z]-[Z_3,X_4] \\ _{4\,e_x} {}_{3\,2}{}_{3\,4} \end{cases}$$

Right-handed [V, ±(m−v)]

$$\begin{gathered} [\pm(V\pm\mu),m] \textbf{ Left-handed} \\ f(r) = [\hat{X}+Y_0,\hat{W}] = [\hat{X},\hat{W}]+[Y_0,\hat{W}] \\ {}_{1\,T}\phantom{=[\hat{X}+Y_0,\hat{W}]=}{}_{0\,-1}{}_{0\,-1} \\ \text{BLUE } i\!\uparrow\leftarrow \text{ v. Anti-BLUE } i'\!\downarrow\to \\ f(s) = [Y_2-W_0,Y] = [Y_2,Y]-[W_0,Y] \\ {}_{3\,C}{}_{+2\,+1}{}_{0\,+1} \\ \text{Anti-GREEN Magenta } \hbar'\!\updownarrow \\ f(y) = [W_4+X_4,W] = [W_4,W]+[X_4,W] \\ {}_{5\,e_y}{}_{4\,3}{}_{4\,3} \end{gathered} \Bigg\} \mu \quad \Bigg\} \dot{m}$$

6.17: Right-Left Lie Fields

Both left and right use the vector as at least part of the change tensor. Right-handed resists change by compounding its static values (m and v). Left-handed compounds change with disorder (μ). The function resolves by tabling u|v(F) on pg. 99 for the Lie derivative definition: $f(F)=[u,v]\to[u,v](F)=$

$$\mathcal{L}_u\, v(F) = \partial_u\, v(F) - \partial_v\, u(F)$$

where:

$$\partial_u\, v(F) = \lim_{t\to\infty} \frac{v(F+tu(F)) - v(F)}{t}$$

6.18: Lie Derivative Definition

This bare coverage of Lie groups and functions does not scratch the surface of spacetime construction occurring in degrees. The next chapters cover the morphology of matter needed to adapt these with the evolving interactions. As our sources illustrate, this line of investigation is significantly involved beyond the scope of this study. We can at least show what you need to get started in that direction.

Matter

Isolated material particles are abstractions, their properties being definable and observable only through their interaction with other systems.

—Niels Bohr, 1934
Atomic Physics and the Description of Nature

Quantum Relativity

Periodic Particle Table

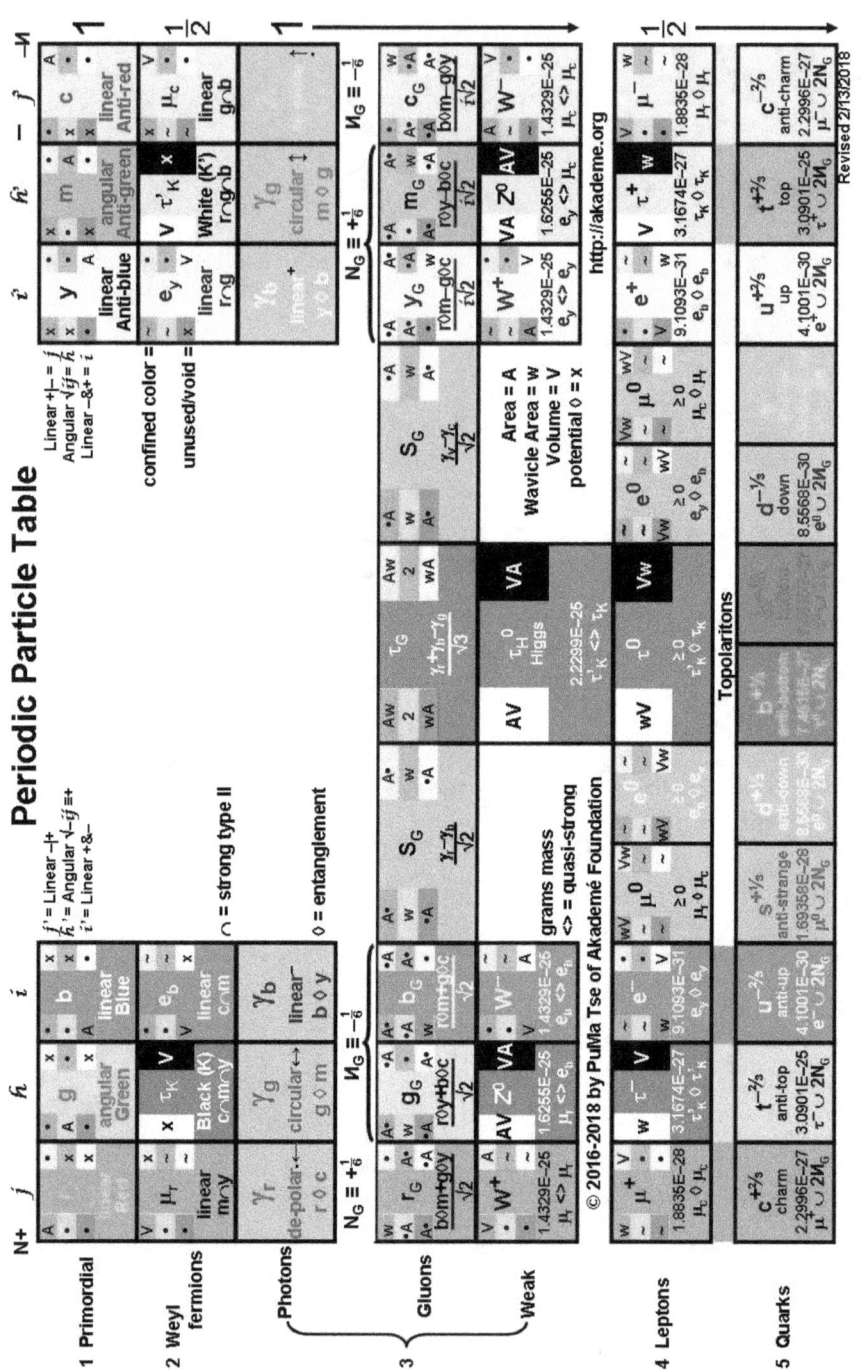

7. Quantum Morphology

At best, physics bundles related groups, and tucks new matter creation under confinement and quantum fluctuation uncertainty.[1] This is due to observational perspective limitations. Biology has already established correlate language and processes for creation and sequential evolution we can apply reconstructing matter from nature's perspective.

Morphology is a study of formation, structures, changes and use of a key component in a field (e.g. word in linguistics,[2] spatial structures in physics,[3] etc.). Morph~ means to change form. Matter comes into form as new (morphogen), evolves (anamorph), compounds (morphotrope), and synanomorphs (multi-forms of the same). Let us begin with sequence.

We have five out of seven types of strong interaction to cover in this chapter. All of these types are particle interactions in the hadronization and isotope processes.

I. Morphogenesis—instantiating new matter. Next chapter.
II. Primordial Bonds—joining 2 colors into type I Weyl fermions and 3 colors into type II Weyl fermions.
III. Primordial Entanglements—forming common identity of microstates by sharing available change spaces as bands (photons and most gluons).
IV. Ambiguous Transitions—4-6 primordials in complex entanglements (gluon-photon singlets) including constriction into temporary volume-surface identities (weak bosons).
V. Mixed Layer Bonding—gluon-photon sets form type II core and surface bonds as flavors form an intermediary mantle (quarks and into hadrons).
VI. Trionic Band Bonding—bonds between available color band edges and surfaces, presumably among trionic band structures to form nuclear isotopes (pg. 149 et seq.).
VII. Degenerate Equilibrium—change information forming levels of equilibrium resulting in degenerate growth and evolution (pg. 172 et

---

[1] Gilman, L. (2018). Virtual Particles. Net Industries. science.jrank.org/pages/7195/Virtual-Particles.html.
Jones, G.T. (2002). The uncertainty principle, virtual particles and real forces. Physics Education. hst-archive.web.cern.ch/archiv/HST2005/bubble_chambers/BCwebsite/articles/06.pdf.
[2] Aronoff, M. & Fudeman, K. (2011). What is Morphology? 2 ed. Wiley-Blackwell. abudira.files.wordpress.com/2014/03/mark-aronoff-kirsten-fudeman-what-is-morphology-fundamentals-of-linguistics-second-edition-2011.pdf.
[3] Mecke, K.R. & Stroyan, D. eds. (2002). Morphology of Condensed Matter: Physics and Geometry of Spatially Complex Systems. Springer.

Quantum Relativity

seq.). As a function of normalizing and sequencing "quantum foam" information, we call this **Wheeler interaction**.

Generations of Matter

The hierarchy of life/biological organization has 12 levels[4] analogous to the hierarchy we call the generations of matter. We are dividing these levels into four groups: pre-cellular (atoms, molecules, macromolecule/bio-molecular complex), organizing (cell, tissue, organ, organ system), ecological (**organism**, population, community, eco-biome), biosphere (the whole).

The pre-cellular group is everything physics would call virtual: primordials (color charges), Weyl fermions (bound colors), and bosons (ambiguous loose interactions). These are so primitive they can only be real (e.g. biological) by being confined (in organic context).[5] Note: macro/bio-molecules are omitted from the image but not the listings.

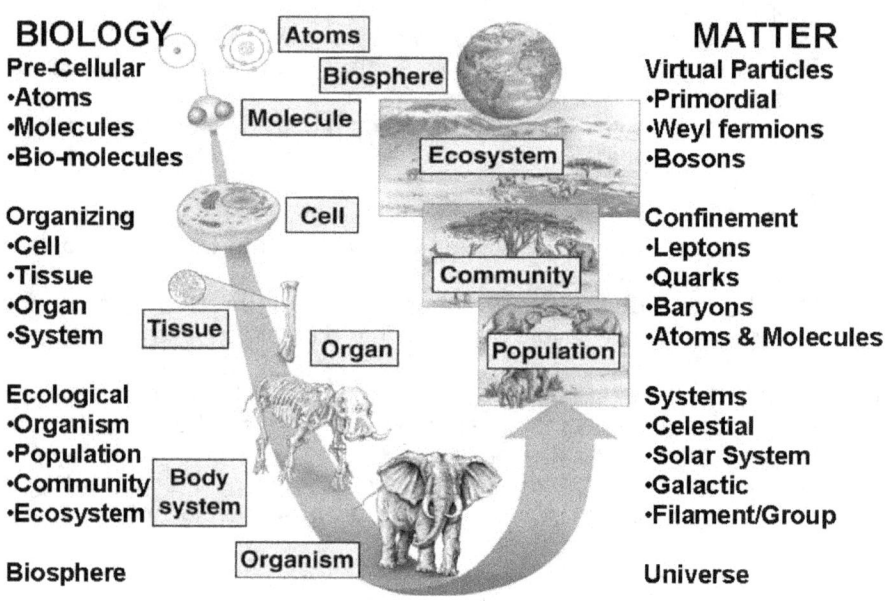

7.1: *Hierarchy of Life v. Generations of Matter*

Confinement is organization. It begins with the simplest that can exist independently (leptons) followed by states dependent (quarks & baryons) on the independently functional (organism is equivalent to atoms and

---

[4] Solomon, E.P.; Berg, L.R.; & Martin, D.W. (2002). Biology. 6 ed. Belmont, CA: Brooks/Cole.
[5] Image (modified): Pen, J. (2013). Introduction to Life Science & Biology. Heartife. https://www.slideshare.net/JuliePen/intro-to-life-science-biology.

molecules of physics). These coordinate into ecosystem levels. You, as a complex organism, are an ecosystem unto yourself.

A population of ecosystems interacts as a community forming a nested system called a biome. Nesting means putting one equivalent inside another. A macromolecule, for example, describes a molecule pattern with potential application. These patterns can be and often are extremely intricate with more member atoms than can be independently counted. They nest and compound.

This speed bump of nesting massive numbers of atom and molecule groups is the same in physics, just in a different point of the modeling. The entire biological model gets squeezed (nested) into only two generations of physics. Stages from macromolecules to biosphere nest into the celestial organism (e.g. planet).

Organisms form system populations in galactic communities, themselves members of a local system of galaxies. The totality of all these is the whole universe—or biologically the biosphere.

The image suggests the hierarchy is cyclic. In effect, the physical state of the biosphere is synonymous with that of the universe of physics generally. High level changes affect everything from the bottom up.

The cycle is completed by redistribution of value (light), which would be the zeroth (0th) generation. It is zero because it isn't actually matter. It is the connection between whole and its most finite and numerous parts.

This connection is "virtual." The parts cannot be examined individually but only together in a confined system. This is also true of the Abstraction Layer of the architecture, making the universe generally virtual.

The universe (biosphere) is first AND last—technically part of the fundamental first group of levels (pre-cellular). From a biological perspective, a change in the weather is a biosphere issue affecting from atoms up the entire hierarchy of levels.

The same is true in physics. Ability to affect everything from the bottom up increases with advance in levels (generations). A local group phenomenon like a gamma ray burst, manifests from the bottom of the process up.

Virtual Particles

Virtual particles baffle many. They look at something like a weak boson and see mass. It must be real if it has mass right? Wrong. Real is also no guarantee of stability. Virtual existence is environment specific and generally occurs as a confined feature in transition—a "transient fluctuation"[6] or disturbance.

Chromodynamics is a Quantum Field Theory (QFT) using additive and subtractive colors as quantum numbers. The colors have non-commutative features making them non-abelian. They are used to describe strong interactions initially among quarks and baryons, mediated by gluons.[7]

---

[6] Thomson, M. (2017). Modern Particle Physics. Just the Facts 101 ©.

Quantum Relativity

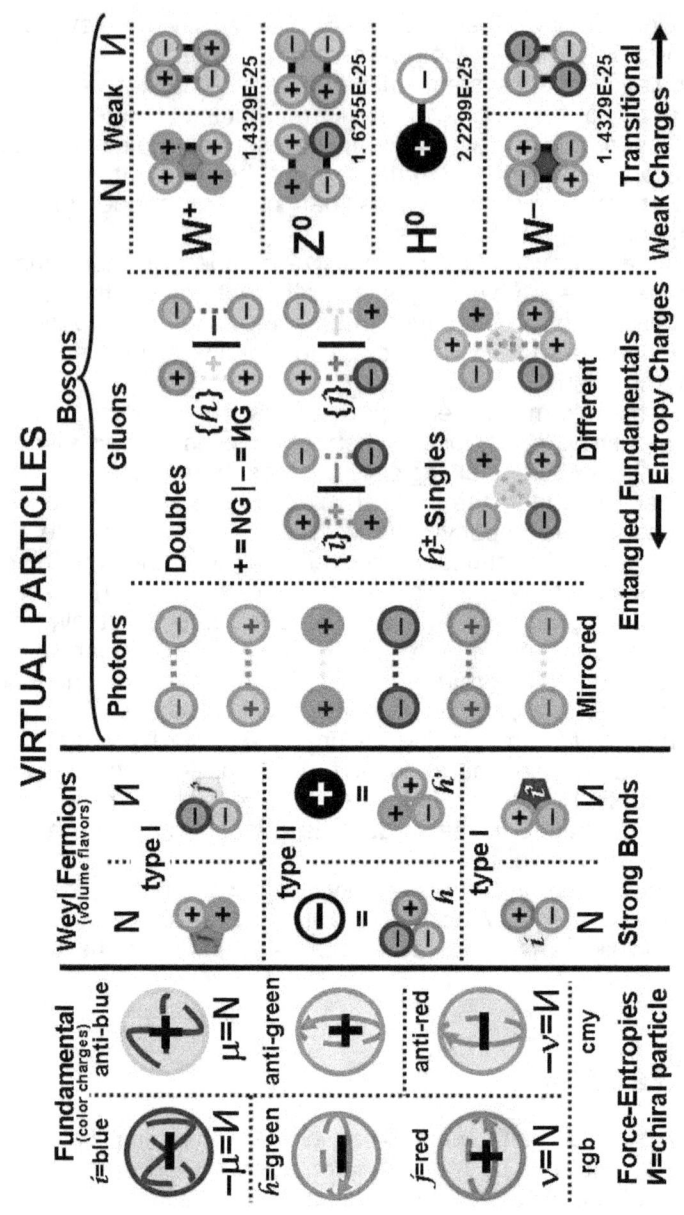

7.2: Virtual Particle Quantum Constructs

---

[7] Manoukian, E.B. (2016). <u>Quantum Field Theory I: Foundations and Abelian and Non-Abelian Gauge Theories (Graduate Texts in Physics)</u>. Springer.

Virtual particles divide into three families: primordials (color charges), Weyl fermions (volume flavors), and bosons (transitional particles). Bosons further divide into photons, gluons, and weak bosons. Of these, primordials only exist, even as virtual, in a relationship—even if that is interacting with their own reflections as photons. They are always mathematical objects.

This diagram is not an indication of ordinary structure. It illustrates content, interaction, and resulting identity. Virtual chirality (sign and color) is an anti-particle analog. It evolves conceptually into helicity (weak sign) and relativistic anti-particles later. Virtual chirality is not an indicator of conventional antimatter. In conventional matter, the chirals work together.

Primordial (Color Charges)

As quantum numbers, the universe does not make much distinction among primordial color charges and gives them no individual existence. Primordial color identities exist strictly as mathematical containers within other identity interactions like microstates. The absence of interaction skips past primordial to annihilation as light (generation 0).

Color identities swap places like a flag blowing in the wind. That analogy is pretty useful because the flag has no control. It is attached to a pole and that determines the limited directions it can blow. The pole would be analogous to a common color identity.

Without the pole, the flag simply blows away. The same is true with color charges. Even though they offer a change function, they need a secondary system to begin grounding them—grounding being planting said flag pole.

One can reasonably argue that primordials never occur, except they do in the most obvious of all places: photons. We can make this argument because a photon is a primordial that divides its value and interacts with its own reflection. Aside from easily modified generalizations of the wave functions, photons are too confined to make color charge distinctions. Take away interaction and primordial is annihilation to generation 0 (light).

Weyl Fermions (Flavor Volumes)

A Wyle fermion consists of the intersection of two or three primordial axes. Such an intersection is a type II strong interaction consistent with part of the strong bonding actions of quarks. It happens when the in-phase volume spaces of the primordials contact. The intersection forms a volume identifiable.

Instead of an intrinsic surface, the interaction provides a color quantum number called flavor as potential for entanglement into a lepton. Conventionally, Weyl fermions are reported as consisting of a volume (flavor) with an attributed Fermi surface—what we have called a negative space. "The Fermi surface is the surface in reciprocal space which separates occupied from unoccupied electron states at zero temperature."[8]

Such a negative space on this level violates relativistic Lorentz symmetry. This is an over-fancy way to explain the volume-surface requirements for the GFE are not adequately satisfied to provide an intrinsic mass value. It is also important to condensed matter physics where Luttinger's theorem shows particle density as a function of Fermi surface-enclosed volume.[9]

"Condensed matter" describes degeneration—the process of matter fully occupying its space. Degeneration re-establishes change function order, such that EMR and EMA values begin to differentiate. An EMA Fermi surface offers minimal restrictions and useful guidance to passing energies. Adding momentum, however, actualizes the GFE.

There are two general designations of Weyl fermions, each providing colored band potentials. We provide a total of six specific particles consistent with resultant flavors of their entanglements (affecting leptons and quarks). We designate all Weyl fermions consisting of only two bound parts as **type I**: rg and cm are electro with i bands; bg and my are mu flavors with j bands.

Weyl fermions consisting of three parts (rgb and cmy) are **type II** (tau flavors) noting their band potentials are quasi-temporal and not color fixed (Heisenberg uncertainty applies). This gives them polarization such that they show different potentials between aspect positions. It also leaves them vulnerable to environmental flavor changes. Most useful of all is a quantum tunneling ability. This allows them to bypass other color variables to act as nucleation and surface for nucleons.

Our type II definition agrees with and expands on: "The type 2 particle exhibits very different responses to electromagnetic fields, being a near perfect conductor in some directions of the field and an insulator in others."[10] Fermions are known to have three flavors.[11] Our Weyl fermion definition shows them as the root of flavor and the subtleties affecting charge.

---

[8] Dugdale, S.B. (Apr. 18, 2016). Life on the edge: a beginner's guide to the Fermi surface. The Royal Swedish Academy of Sciences. iopscience.iop.org/article/10.1088/0031-8949/91/5/053009.

[9] Halboth, C.J. & Metzner, W. (1997). Fermi surface of the 2D Hubbard model at weak coupling. Springer. https://link.springer.com/article/10.1007/s002570050318.

[10] (Nov. 25, 2015). 'Material universe' yields surprising new particle. Princeton University. https://phys.org/news/2015-11-material-universe-yields-particle.html.

[11] (May 23, 2017). Weyl fermions exhibit paradoxical behavior. Leiden Institute of Physics. https://phys.org/news/2017-05-weyl-fermions-paradoxical-behavior.html.

Unbound Identities

Type I strong interaction is instantiation (next chapter). Type II strong interaction is bonding to form Weyl fermions. Type III and IV strong interactions are entanglements and the transitional weak bosons.

Types III & IV are not firmly bound like the axis conjunction of sharing phase spaces in type II. They start with what are commonly called entanglements. An entanglement at this level is basically a pseudo-strong bond. It occurs at the edge of the "out of phase" surface and axis space.

Generation-wise, the result is two generations above the parts. Entangled (strong type III) primordials are bosons. Weak bosons specifically are between the edge and in-phase positions—in presumable transition to bound and entangled. They are given the separate designation of type IV strong interaction because their microstates simultaneously provide surface and volume resolutions. Type III just adds a surface.

Photons

The only clear distinction between primordial, light, and photon is focus. To create a photon, a primordial interacts with itself by dividing its resources in three more compact spaces. It can now use these spaces to play microstates hot potato with the never-resting light value. Renormalizing into temporal effect, especially degeneracy, dilates and reduces the need for this quantum juggling act.

Light-primordial-photon all have the same wave qualities that can be used to generalize color identity. The moment you plant this flag someplace, it becomes less generalized. It generalizes because the helical (mirrored) quality of the parts require synchronous (neutralizing) bands.

The change function of any interaction is antithetical to the interacting parts. A direct chiral (red-cyan) neutralizes the change function and annihilates. That leaves magenta and yellow as interactive potentials for red to participate in.

When the parts are identical, as with photons and charged leptons, the parts rotate to perfectly mirror each other. Red on red provides either magenta or yellow, but the rotation shows as r–r with g:m|b:y bands. These alternate in microstates. As a time-neutral confining surface, they mask and impede color specifics and interactions.[12] Photon adaptability makes them perfect building blocks in material evolution.

As in all such cases, time neutral creates an environmental susceptibility and opportunities. A lens filter easily changes color identity shaping the light. Indeterminate color is a quantum tunneling way of getting around other color interactions. This makes quark anatomy ambiguous, but simplifies topolaritons.

---

[12] Hansen, L. (Feb. 27, 1997). The Color Force. webhome.phy.duke.edu/~kolena/modern/hansen.html.

Quantum Relativity

The rotation affects synchronicity (see pg. 95) differently for simple band pairs versus flux tubes. A flux tube consists of a band pair and the pairs are separated. Photons are synchronous, where a similar muon's (my–my) flux tubes are asynchronous. The reason is that each primordial is offering band potential. Each rgb|cmy offers three band potentials—they just aren't simultaneously active.

Gluons

We simplified color-anti-color to additive-subtractive for accessibility purposes on many levels from keyboard characters to actual coloring. While we still use the same definitions,[13] we simplify them into collective identities and provide ◊ to indicate swap uncertainty.

We add a weak charge equivalence layer to the definitions to aide in understanding quark charges. When alternating with a photon, the photon also has an equivalent attributed ±1/6 weak charge. This way, two gluons or a gluon and photon combine with a lepton to make a quark.

There are eight gluons, but we separate the gluon "singlets" as ambiguous like the weak bosons. The six basic gluons are simply entangled color and anti-color of different types that swap color/anti-color roles but retain the same confining entanglement surface identity.

All gluons are strictly surfaces entangled with other surfaces. They have no mass because there is no volume for a surface to interact with. Adding energy into them expands and rotates their bands. Sudden discharge of that energy is called snapping.

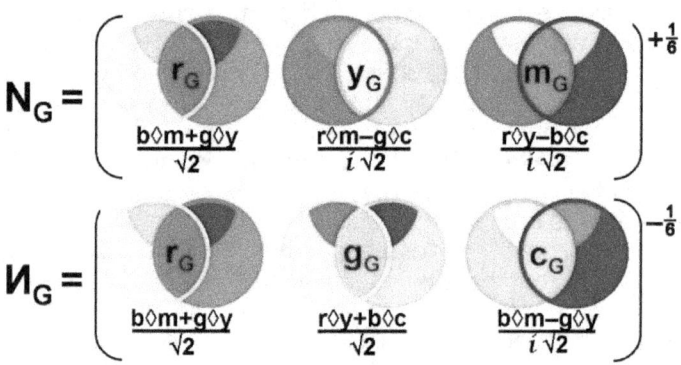

7.3: Confined Color to Weak Charge Equivalence

Energy can quantize resulting, the snap splitting into two new gluons, each gluon having half the parent's parts—analogous to quarks.[14] An r_g

---

[13] Griffiths, D. (1987). Introduction to Elementary Particles. John Wiley & Sons. pg. 280. ia800909.us.archive.org/5/items/GriffithsD.Introductionto elementaryparticlesWiley1987/GriffithsD.Introductiontoelementaryparticles %28Wiley%2C1987%29.pdf

gluon has gy or bm parts. Assuming either gy or bm at division provides the same offspring: y_g and m_g. This is a form of instantiation (morphogenesis) we term mitosis—the offspring have the same scale and number of parts as the parent.

Ambiguous Cases

There are two types of ambiguous cases, both of which involved complexes of four or more entangled primordials. Gluon-photon singlets are equally entangled red-green, or red-green-blue photons. All gluon and photon interactions apply, but the bands do not apply simultaneously.

The singlets are out of phase strong type III entanglement interactions. When their phase volumes interact, they become weak bosons. Conversely, weak bosons devolve into distinct gluons.[15] Which weak boson depends on the specifics, which is a key difference between type III and IV interactions.

Unlike the primordial entanglement, the type IV interaction limits color axes. This is a required level of stability to establish volume even temporarily. Z bosons juggle rg:bg volume:surfaces together to neutral, just as the Higgs does rgb:cmy. For charges, W bosons do rg OR bg separately.

Despite similarities, only the rgb singlet stands a chance of direct translation. To reiterate, the key differences between singlets and weak bosons are:

- Singlet part interactions entirely occur out of phase range (not allowing volume)—gluon/photon primordials.
- Weak boson parts interact in the intermediate zone between in phase and the edge of out of phase—having simultaneous qualities of both primordials AND Weyl fermions.
- Singlets utilize all their parts and interactions where weak bosons limit their parts and interactions.

Weak Confinement

Weak bosons provide a basic template upon which confinement builds. Key to this template is the emergence of Fermi surfaces (negative spaces) resulting from optimizing the use of space. This evolves chirality into the concepts of helicity (weak charge) and anti-matter.

---

[14] Butterworth, J. (July 18, 2010). Quarks, Gluons and Jets. lifeandphysics.com/2010/07/18/quarks-gluons-and-jets/.
[15] Djouadi, A. (Dec. 1997). Decays of the Higgs Bosons. Université Montpellier. cds.cern.ch/record/340786/files/9712334.pdf.

Quantum Relativity

The interactions responsible are among parts of different scales and their composites of different magnitudes. The complex interactions force differentiation of quantum local from relativistic group fields. Only where these satisfy and sustain GFE requirements do relativistic qualities like Fermi surfaces, momentum, and mass properly apply.

Gluon and photon bands are the same magnitude as their bands. All further bands and flux tubes, no matter how many, are of that "primitive" magnitude. Primitive here does not mean small, but rather simple. Greater simplicity has greater energy and spatial requirements.

Charge potential depends on the consistency of band pairings. A simple pair (gluon) has opposite values in like outward positions. This gives a color charge. Photons twist these to neutral temporal.

Flux tubes have inner and outer sides in hypercomplex temporal conditions. The color state of the outer sides defines the charge. An electron, for example, has outer $-i$-bands, a positron has $+i$-bands, and the e-neutrino twists these. Weak boson surfaces emulate flux tubes.

Confining is a process of achieving stable units: quantization. Quarks and sub-quark topolaritons are intermediary particles. Most particles are to some degree intermediary. Stability requires the highest possible degree of order toward singularity. Red and cyan are the most ordered color fields. The next determining factor of stability is complement: the ideal charge state. Leptons are the first relativistically relevant charge states.

Leptons

Each lepton consists of an entangled pair of related Weyl fermions (flavor volumes). Neutrinos are a simple chiral pair causing their flux tubes (band pairs) to twist over themselves. This is the flux tube equivalent of synchronous bands as we see with photon bands. In both cases, the twisting bands are quasi-temporal and neutralizing.

Without momentum, synchronicity deprives neutrinos of relativistic qualities. Inconsistent microstate distribution of value among volume and surface also limits relativistic qualities. Unlike photons, lepton motion is relativistic, so part of the observed mass is intrinsic or applied momentum. Note: we prefer the particle table (pg. 104) so no one gets confused that the particle breakdowns can be viewed as literal geometries.

After neutrinos, deriving charge from chiral flavor source is more complex. Primordial colors merge in each Weyl fermion to the equivalent flavor concept. The change function associated with the color surface is now evaluated as a volume because the change axes share common geometric origin. Volume is the result, and the color assigned to the flavor is its potential. The first field sign in the color's change function is the lepton charge ($\check{\imath}, \check{\jmath}, \check{h}$=+; i, j, h=−). The j and $\check{\jmath}$ bands cross over.

Quantum Relativity

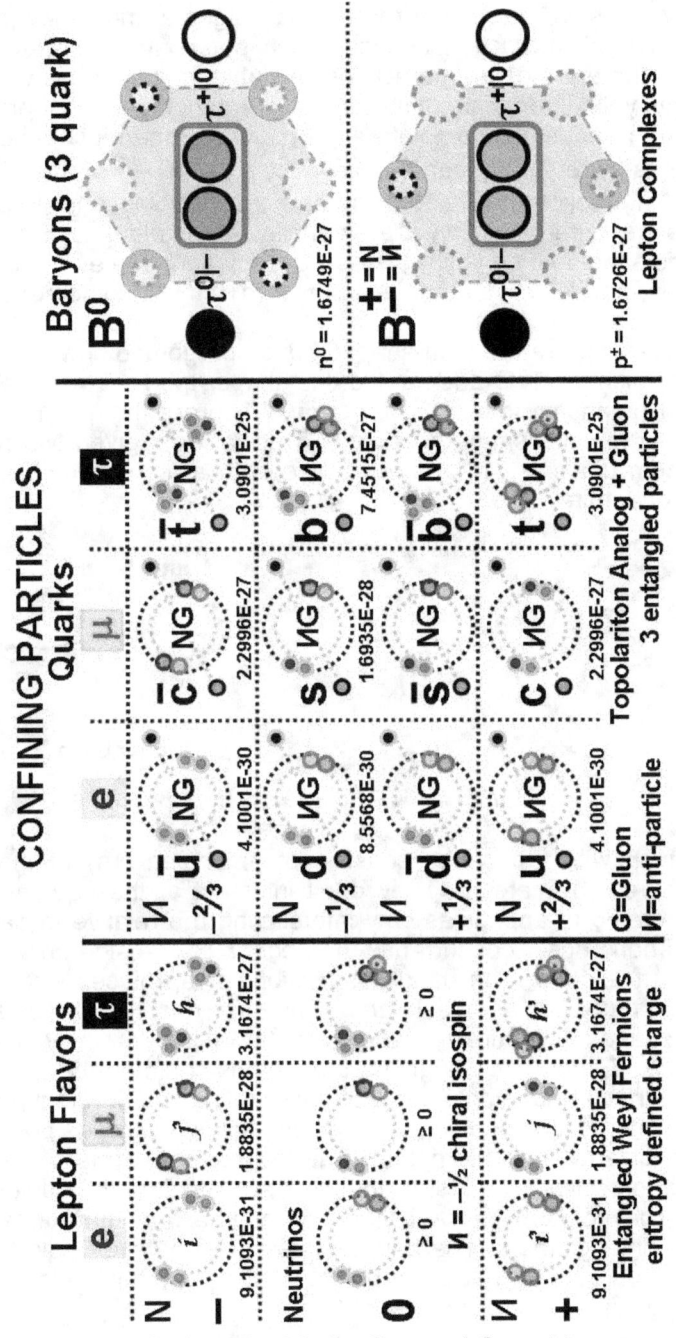

7.4: Confining Particles Pre-Structural Compositions

Quantum Relativity

With gluons we discussed red having magenta and yellow potentials. The subsequent interaction defines which potential actualizes. This can alternate if there are other interactions available. In a bond, as with red and green, only yellow remains available. Each entangled identity offers a pair of yellow ($\overline{i}$) bands. To complete the interaction, one must either have or fake i bands by rotating (equatorial) aspect.

Color interactions strive for cmy and rgb combinations. Even though opposites are chiral, they can rotate to imitate each other, as with e⁻=rg–rg. The bands should all be yellow, except the rotation creates a pseudo-blue. Conversely, an electro-neutrino is additive rg+cm with synchronous bands (see pg. 95).

More baffling are the trion-shaped bands of rgb and cmy (Type II Weyl fermions). Michael R. Evans named this shape "trion re" for its shape and the crystalline prism effects it has on light. Of course the quantum universe couldn't just leave a shape so simple. True to form, given two potentials, the quantum universe is happy to utilize both at its convenience. It is a quantum quagmire.

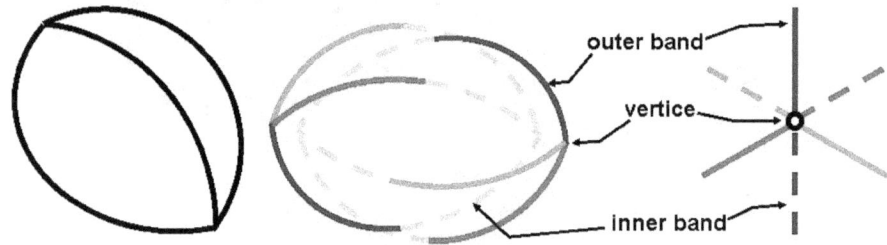

7.5: *Michael Evan's Trion and Trionic Bands*

The trion with rgb bands consists of entangled cmy at the vertices (ends) where rgb intersect. They don't intersect at the equator because they are rotated to appear as cmy intersecting rgb relative to each other. They also don't occur concurrently, and edges of one side are connecting to faces of the other. The band associations swap places since they are not fixed commodities and each interacting primordial has two potentials.

This sounds like purely academic detail. It is extremely relevant on many levels. Without the quagmire, tau leptons could not be used as both core and surface of baryons. It is also extremely relevant with type VI (6) of the strong interactions bonding baryons and making isotopes possible.

Their bizarre quasi-temporal nature makes exposed tau bands vulnerable to oscillation,[16] identity change, and ideal for their unique role in hadrons. That role can easily be described as a quantum tunneling-style system of entanglement. The bands in use are whatever (color) quantum number is available at that cycle point. Similar opportunism is common in neutrinos, gluon singlets, and weak bosons.

---

[16] Affects mass $\Delta m^2 L/E$. Boyd, S. (Mar. 24, 2015). <u>Neutrino Oscillations</u>. https://warwick.ac.uk/fac/sci/physics/staff/academic/boyd/stuff/lec_oscillations.pdf.

Quarks

Quantum number convenience is the name of the quark game. Each quark is flavored by a lepton (e, μ, τ) providing base charge. This is populated by a gluon-photon pair alternately interacting as two gluons modifying the base charge. A photon-lepton combination is known as a topological polariton (topolariton).[17]

Gluons and photons do not have weak charges, unless they are injected in an interaction creating a weak field. Then their individual effect is 1/6—likely due to swapping with two gluons. As with the leptons above, the first sign of the entangling change function is the sign for the gluon. The photon follows the gluon's suit, which means its primordial parts begin with the opposite sign.

The gluon-photon pair have equal charges in the same direction affecting the charge by ±⅓. Down (e), Strange (μ), and bottom (τ) quarks are flavored by a neutrino. The gluon-photon pair gives them 0±⅓ charge. Charm, up, and top quarks are flavored by a charged lepton. The gluon-photon charge effect is ±(1–⅓)=± ⅔.

Hadronization

Quark bonding is more like the nested meshing of a novelty puzzle ball as below. It is really better described as a lepton combination than a quark combination. We are terming it strong type V (five) because it has unique elements. Like the puzzle ball, parts from the surface also make up the core, while others simply mesh like a layer of filler.

7.6: IQ Puzzle Ball/Quark Bonding Example

From a dissection perspective, this complex system is too compact and offers too many ways to break down. To describe its formation is

---

[17] Refael, G. et al. (Jul. 1, 2015). Topological Polaritons. CalTech. https://journals.aps.org/prx/abstract/10.1103/PhysRevX.5.031001.
[18] (2018). 3 IQ Puzzle Ball. Littleton, CO: Dollar Item Direct. dollaritemdirect.com/iq-puzzle-ball.aspx.

Quantum Relativity

simply strong bonding is to not give the full set of responsible interactions credit. This is really degeneration, which is our last section.

The quark's identity is almost completely lost. The gluon-photon sets alternately form type II bonds—more Weyl fermion components. These come in tightly wound layers and cross-layers of flux tube band pairs. At the core of a hadron, for example, is a degenerated Higgs boson. It is degenerate because it is only its volume.

Gluon-photon bands in quarks already conditionally passed through a lepton layer, separating their halves between outer and inner. For hadrons they now form tau bands holding an outer-core of Weyl fermions and their mantle of lepton-like interactions between outer membrane and core volume (both unavailable).

Membrane includes volume and flux tube bands available as Fermi surface. Membranes can weakly interact and have strong type VI bonding potential. This numbering is convenient for remembering that type V occurs between quarks of generation 5 and type VI for baryons of generation 6.

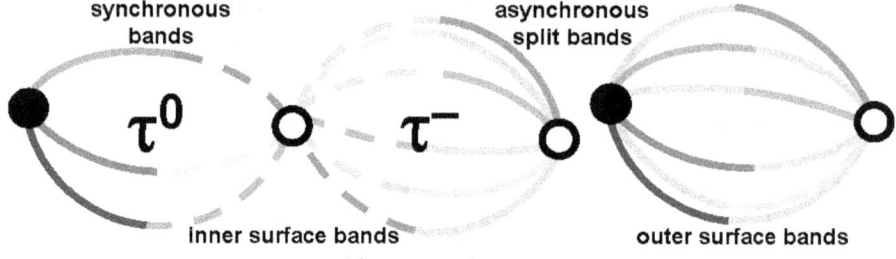

7.7: Trionic Band Synchronicity

The separated surface is very important. It deprives the Higgs core from a direct GFE interaction leaving it a quantum variable in this respect. The leptons that were quark flavors now make up the outer core and mantle layers of the baryon. This layer is another membrane: a surface with depth. Baryons thus have two membranes and a volume filled in part or completely (as with nucleons).

Dilation has a profound impact on how the fields of a degenerate like a baryon behave, and how the spaces are being defined. They no longer behave anything like particles of lesser generations. For this reason, we will pause to complete the rest of our particle discussion before delving into degenerate details.

8. Instantiation

Every story has a beginning, middle, and end. The quantum universe is a setting, not a story. Settings are timeless unless you happen to be the weary traveler stuck taking the long way through the pages of a story.

In physics, matter is the plotline. It determines beginning, evolution, and end. The story is a unique combination of substance in motion changing to its local conditions. The story is evolution. Evolution can play any role in the existence of matter from creation to termination—arguably every role.

Using story elements invites ethical dialogs like: What is the purpose of ____? Science and the physical universe simply apply definitions in context. Being a definition in context plays a role in transformations and evolution on many levels. These roles can be active, passive, giving, or receiving, like being redefined by local turbulence.[1]

Strong Type I

The four reproductive phase-stages of mitsosis/meiosis[2] outline the first type of strong interaction: creation of new matter at any scale. This is a presumably passive process of quantum fluctuation associated with Heisenberg uncertainty and confinement.[3] The particle or otherwise available spacetime is acted upon, accumulates, reacts, and diverges into multiple forms (cladogenesis[4]).

The first two stages provide explicit details of interphase—routine existence. Each consists of a sequence of sub-processes. The references provide a better outline of the cellular analogs that do not always convert directly into the new matter process. Life, mind and civilization are imperfect mirrors attempting to emulate the perfection of physics.

---

[1] Richings, R.J. & Faucher-Giguère, C.A. (Nov. 23, 2017). <u>The origin of fast molecular outflows in quasars: molecule formation in AGN-driven galactic winds</u>. Oxford Academic. academic.oup.com/mnras/article-abstract/474/3/ 3673/4655190.

[2] Hartwell, L.H.; et al. (2008). <u>Genetics From Genes to Genomes</u>. New York: McGraw-Hill.

[3] Gilman, L. (2018). <u>Virtual Particles</u>. Net Industries. science.jrank.org/ pages/ 7195/Virtual-Particles.html.

Jones, G.T. (2002). <u>The uncertainty principle, virtual particles and real forces</u>. Physics Education. hst-archive.web.cern.ch/archiv/HST2005/ bubble_chambers/ BCwebsite/articles/06.pdf.

[4] McClean P. (1997). <u>Population and Evolutionary Genetics: Speciation</u>. ndsu.edu/ pubweb/~mcclean/plsc431/popgen/popgen6.htm.

Prophase establishes the reproductive potentials (leptotene, zygotene, pachytene, diplotene, diakinesis[5])
Condense—available focalizing spacetime, like color charged attributed value spaces (bands);
Conjugate—paired groupings or topological ordering;
Connect—pair mirroring or symmetry;
Crossover—energy connecting exchange like microstates and oscillation;
Confine—Fermi surface encapsulation.

Metaphase—spindling[6]/value fills a confining change shape:
Adjacency—capture/focus of value;
Configuration—information mirroring, distribution, and smoothing;
Carry/Accumulation—energy not lost compounds with new adjacency up to…

Anaphase—conception/establishment of new (sister parts separate[7]):
Quantization—establish unit identity proportional to contextual availability;
Snap—attributed space separates from acquired value;
Rip/Unzip—radiant value divides by absorption into chiral units;
Alignment—initial interactions of new parts.

Telophase—differentiation/separation into distinct identities[8]:
Cycle initiation—microstate cycle of new parts and interactions;
Re-alignment—interactions and chirality of parts established;
Conflict resolution—quantum number swapping/leaping to tunnel or jet through exclusion conflicts;
Confinement—into prophase/interphase.

It is easy enough to envision perturbations of virtual particles forming and evolving through this process in the band spaces of particles. Imagining this in the complex field spaces of a galaxy developing confined and system matter is a bit harder. It seems further complicated by having not one but two types of field generators (cyan and red singularities). Shape is the only significant difference.

The image below illustrates the smooth generic field condition of the universe acting on a red singularity. There are color-predisposition spaces

---

[5] Staveley, B. (2017). <u>Principles of Cell Biology</u>. Memorial University of Newfoundland. mun.ca/biology/desmid/brian/BIOL2060/BIOL2060-20/CB20.html.
[6] Kimball, J. (Apr. 5, 2014). <u>The Centrosome</u>. biology-pages.info/C/Centrioles.html.
[7] Kimball, J. (Apr. 5, 2014). <u>The Cell Cycle</u>. biology-pages.info/C/CellCycle.html.
[8] Miko, I. ed. et al. (2014). <u>Telophase</u>. nature.com/scitable/definition/telophase-128.

in the field where x-ray (+i) and gamma ray (−i) emissions regularly occur. There are vast spaces where seemingly nothing occurs (negative/unused space). Around these is a toroidal feedback system.

The universe of things imperfectly acts on the singularity to generate a toroidal field. We must remember that the universe of things is not just local things but also the myriad of overlapping singularity-horizon bubbles creating large scale anomalies. While a member of the electromagnetic family, this is no ordinary magnetic field. Ordinary magnetic fields also occur, but they are not as effective at steering the direction of matter let alone light.

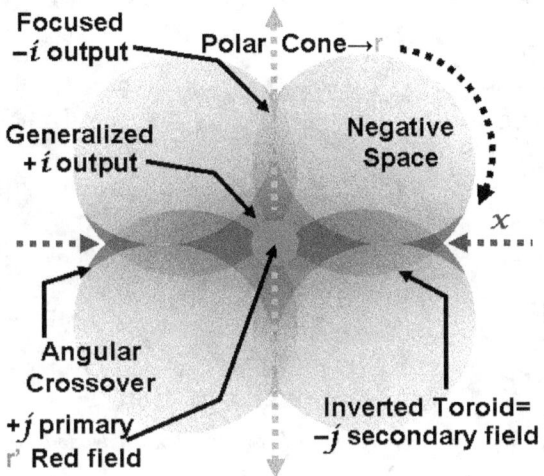

8.1: Complex Toroidal-Phase-Interactive Singularity Fields

You cannot see a beam of light from the side unless it is reflected off something and comes straight into your eye. The light caught in this field also cannot be observed without it reflecting off something, like the light reflecting off the rings of sombrero galaxies and gas clouds on galactic rims. This isn't just stellar light reflection.

Despite the illustration not accounting for system information or spacetime anomalies, there are clear focal points and dilations. At these points, energy and/or matter focus resulting in new and evolving matter from virtual photons to solar systems and everything between. Galaxies literally create everything from seemingly nothing.

Initially the information is negligible allowing the creation of hydrogen (Lyman-α=1216Å wavelength) evolving from simple emission to stellar break.[9] When we add in system information, the perturbations become complex dynamics. Information evolves in these dynamic systems as the complexity of matter evolves.

---

[9] (2012). <u>Lyman-Alpha and Lyman-Break Galaxies</u> Hubble/ESA and NASA.   hubblesite.org/hubble_discoveries/science_year_in_review/pdf/ 2012/exploring_lyman_alpha_and_lyman_break_galaxies.pdf.

Cladomorphology

A clade is a group classified by common ancestry.[10] This term is problematic in physics where more than one path to the same outcome is expected. The plot paths of change are more easily classified. It is in these contextual paths that matter stories emerge, unfold, evolve, and end.

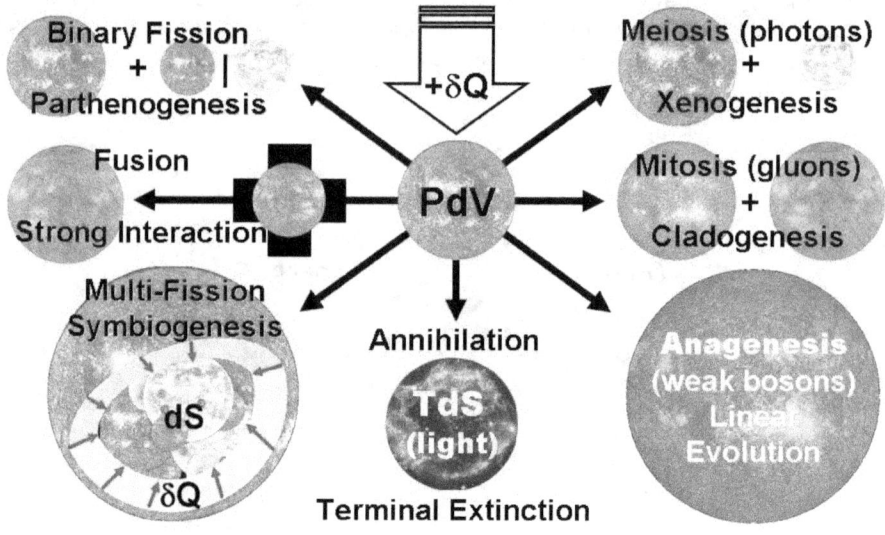

8.2: Cladomorphology Types and Thermodynamics

Darwin's concept of evolution of the fittest suggested a gradual system of adapting to the environment. This led anthropologists to seek a clean progression in human evolution from primates. The concept has a limited degree of truth. Selective breeding shows dramatic changes happen from one generation to the next with a significant and consistent behavior change, like domestication.[11]

In thermodynamics, enthalpy (H) describes the total energy of a system in a change of state (e.g. solid-liquid-gas). It is a function of heat (E) and linear pressure (P) applied to a volume (V): H=E+PV.[12] The same concept is true for biological evolution in that change in behavior (E) is combined with the motivation (P) of a population (V).

As a change boundary condition, enthalpy can be thought of as a quantization value. This can be triggered via enthalpy equivalent relativistic

---

[10] Polly, P.D. (2013). Phylogenetic definitions of taxonomic groups. indiana.edu/~g404/Phylogenetic%20definitions%20of%20taxonomic%20groups.pdf.

[11] Trut, L.N. (Mar.-Apr. 1999). Early Canid Domestication: The Farm-Fox Experiment. American Scientist, Vol. 87.

[12] Hall, N. (May 5, 2015). Enthalpy. grc.nasa.gov/www/k-12/airplane/enthalpy.html.

momentum. For particles, the behavior change appears in oscillation—periodic motion consistent with identity.[13]

Oscillation is a product of intrinsic information and is subject to normalization conflicts with transient information. The longer transient energy remains, the more their separate information normalize. This consistency and other environmental pressures increase the probability of change at and beyond enthalpy as shown in neutrino studies.[14]

Enthalpy is the trigger for change when thermodynamic energy (δQ) applies to a contextual (PdV) path. We assume the path is material, but it can be a field or incidental focus (perturbation), creating a loophole in the material requirement for creating new matter. There is always more than one way to twist the plot.

Here, energy is vaguely defined as a capacity for work (δW).[15] Mass contains capacity and resists work. Time contains and resists changes where work applies. The work here is evolution or creation of matter. We will call the creation of a new matter object **instantiation** to include more than just virtual perturbations.

Instantiation—The contextual process of bootstrapping a material identity into existence; the creation of an actual object instance in a context (object) oriented system[16].

Instantiation establishes an active change function generally in the space of an inactive change function (e.g. entanglement band). Another is signal interference, which is a temporary perturbation—a virtual instantiation. Instantiation is opposite to annihilation where a change function is lost with value evaporating as light.[17]

The assumption of instantiation is spontaneity. An identity is called into or out of existence like a virtual entity. Such spontaneity is generally classified under quantum fluctuation and Heisenberg uncertainty.[18] Virtual particles (primordials, Weyl fermions, and bosons) are always instantiated—commonly by mitosis and meiosis (later).

Singularities are virtual identities, as is the case of extreme degeneracy (extreme pressure-density[19]). Both are generally a byproduct

---

[13] Evans, L. (2010). Oscillations. webhome.phy.duke.edu/~lee/P53/osc.pdf.
[14] Kayser, B. (2011) Neutrino Oscillation Physics. Batavia, IL: Fermilab. hep.wisc .edu/~sheaff/PASI2012/lectures/BorisK-Oscillation.pdf.
[15] Nave, C.R. (2017). Work Energy Power. Georgia State University. hyperphysics .phy-astr.gsu.edu/hbase/work.html.
[16] Rouse, M. & Macleod, J. (Sep. 2005). WhatIs.com: Instantiation. whatis .techtarget.com/definition/instantiation.
[17] Strassler, M. (Mar. 25, 2012). Particle / Anti-particle Annihilation. profmattstrassler.com/articles-and-posts/particle-physics-basics/particleanti-particle -annihilation/.
[18] Župančič, A.O. (Jul. 17, 1965). Creation Rate of Matter and the Heisenberg Uncertainty Principle. Nature: 207, page 279.
[19] Carroll, B.W. & Ostlie, D.A. (2006). An Introduction to Modern Astrophysics. 2 ed. London, UK: Pearson.

(stellar remnant[20]) of parthenogenesis (binary fission) splitting an existing identity[21].

Confined particles and even degenerate matter also only require perturbation of energy in the right spacetime density conditions. While hard for us to imitate, the complex fields of galaxy-singularity interactions can mass produce protons and neutrons (e.g. Lyman-α). New matter is generally produced in confined complex systems making direct observation or imitation extremely difficult.

Neutrino synthesis is a common example in the very familiar stellar fusion process.[22] Such complex systems break down (fission) and interactively create not one but many types of matter. The interactions form cooperative fission-fusion cycles into hadronization and nucleosynthesis.[23] Such cooperative evolution is symbiogenesis.[24]

Anagenesis is adaptation in isolation.[25] A familiar example among particles is the part of nuclear fission where a neutron becomes a proton. Another is stellar evolution. Systems evolve in the composition of their parts, like stars evolving toward heavy elements. Some of these changes are simply adaptation to energy conditions. Others are from information equilibrium interactions among the parts of a complex. Anagensis also occurs in cyclic evolutionary processes, like the fusion process in stars.

Oscillation is the common physics synonym for anagenesis due to intrinsic behavior change. Usually it is one identity transforming into another of the same class, like leptons and quarks changing flavors, photon filtering and gluon color changes. We can call these horizontal anagenic changes. A lepton transforming into a photon[26] would be vertical.

---

[20] Sulehria, F. (2005). Stellar Remnants. novacelestia.com/space_art_stars/ stellar_remnants.html.
[21] (2018). Binary Fission. Cornell University. micro.cornell.edu/research/epulopiscium/binary-fission-and-other-forms-reproduction-bacteria.
[22] Larson, K. (2006). Neutrinos! University of Wisconsin, Madison. astro.wisc.edu/~larson/Webpage/neutrinos.html.
[23] Terzian, Y. & Herter, T. (Oct. 17, 2012). Stellar Energy and Nucleosynthesis. Cornell University. astro.cornell.edu/academics/courses/astro1101/lectures/13StellarEnergyNucleosynth.pdf.
[24] Margulis, L. (1981). Symbiosis in Cell Evolution. San Francisco: W.H. Freeman.
[25] McClean P. (1997). Population and Evolutionary Genetics: Speciation. ndsu.edu/ pubweb/~mcclean/plsc431/popgen/popgen6.htm.
[26] Minkel, J.R. (Jul. 22, 2002). Two Photons Diverged. Phys. Rev. Focus 10, 3, https://physics.aps.org/story/v10/st3.

Baryogenic Asymmetry

The baryogenic asymmetry problem basically asks why atoms and the emphasis on matter over antimatter? Why do we see matter and antimatter created together but in the end see so little antimatter? In 1967, Sakharov suggested thermal equilibrium, Baryon number, and charge-parity (CP) symmetry violation.[27]

The study of new matter generally focuses on quarks undergoing mitosis splitting into quark and anti-quark pairs. It is low-energy and easy to observe. This is ONLY ONE way quarks can be created. Quarks were (and often still are) described as fundamental. We've seen they are far from it. They aren't even necessary to create baryons.

Baryogenic asymmetry is normal because the universe wasn't constructed by the same rules as our limited ability to observe it. Let us recall that chirality is a feature of virtual particles which are all more fundamental than quarks. Chiral conflicting parts are common such as the Higgs, neutrinos, and all the quarks deriving from neutrinos.

Chiral components strategically cooperate and confine into relativistic matter. The concept of helicity evolves from and replaces chirality (see pg. 48 et seq.). Virtual particles are familiarly created with their antiparticles, but the definition of antimatter changes with confinement. Being born together doesn't mean they die together. The familiar path is not the only path to perturbation, and not all paths require pair creation.

Helicity is relativistic tying in with high-level phase conditions. The universe is left-handed. Left handed means rotation and trajectory go opposite directions—naturally tearing the identity apart. The universe draws out disorder (μ) following the second law of Thermodynamics. Value available for disorder is systematically lost. The parts are only relevant to finding the most efficiently ordered definition (stability).

More pressing than the baryon asymmetry is electro-flavor symmetry (electrons, down and up quarks, neutrons and protons). These are all based in electro (y|b) type I Weyl fermions consisting of red-green (rg|cm) ordered combinations. The y|b designation translates into Fermi surfaces. This is an ideal combination of having right-handed volume providing a null left-handed interface with the left handed universe.

The alternatives to electro-flavor are mu and tau. Charged muons and tauons have average life expectancies: $\mu^-=2.197\times10^{-6}$ and $\tau^-=2.906\times10^{-1}$ seconds.[28] These reflect their disposition to disorder versus the stable electro-flavors and neutrinos.

---

[27] Sakharov, A.D. (1967). Violation of CP invariance, C asymmetry, and baryon asymmetry of the universe. Journal of Experimental and Theoretical Physics Letters. 5: 24–27.

[28] Amsler, C. (2008). Particle Data Group. pdg.lbl.gov/2008/listings/s035.pdf.

Quantum Relativity

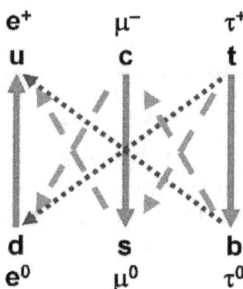

8.3: Quark Decay Path with Flavors Indicated

Primary decay paths are in red, secondary in dashed grey, and least in dotted blue. This follows the Cabibbo–Kobayashi–Maskawa (CKM) matrix.[29] Quarks decay from charged μ (charm) and τ (top) to stable neutrino (strange and bottom).

Stable electros, however, decay from neutrino (down) to +charged (up). +Up has right-hand oriented volume and surface. Baryogenic asymmetry is expected when virtual chirality pairing is not confused with helicity decays toward right-handedness.

Mitosis v. Meiosis

The words "new matter" imply the spontaneous creation of something from nothing—or at least completely unlike things. All matter is essentially energy value applied to change in a space ($m=E/c^2$). This is our list of unlike "nothings" that are actually somethings: value, change, and space.

The creation of something from nothing then familiarly follows one of two general paths: mitosis producing twins with the same number of genes, and meiosis quadruplets reducing the genes.[30] We will start with the particle version and double back later for grander scale productions. Among particles:

Mitosis—The creation of two new material identities with the same entropy but of opposite states. When formed, the new material identities entangle or bond separately with the parent particles (e.g. gluons and quarks).

While this is known to occur among gluons, confinement makes quarks observationally accessible. As energy is added to bands, the bands rotate and expand until they snap at new identity formation. Alternatively, the pressure of collision can convert into workable energy to create new matter fitting the available change conditions of the colliding parts.

---

[29] Ceccucci, A., Ligeti, Z., & Sakai, Y. (Feb. 2014). 12. The CKM Quark-mixing Matrix. http://pdg.lbl.gov/2014/reviews/rpp2014-rev-ckm-matrix.pdf.
[30] Scott, S. et al. (May 17, 2017). Mitoses versus meiosis. yourgenome.org/facts/mitosis-versus-meiosis.

q
q̄
q
q̄
q
q̄
31

Snap!
32

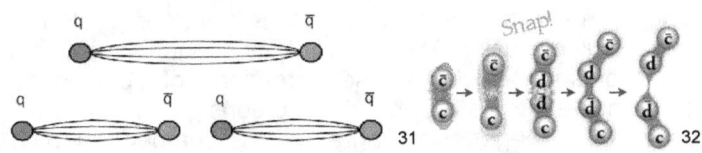

8.4: Band Growth to Mitotic Snap

These snapping illustrations are better suited for gluon mitosis, with energy simply accumulating in the bands. Instead of quark-antiquark it should be color-other anti-color like red magenta (r+m). This has b-y bands such that at band snapping you get r+y and b+m gluons.

Meiosis—The creation of two or more new material identities generally of fewer change features. This would include photon emission from electron quantum leaping, and jet emissions.

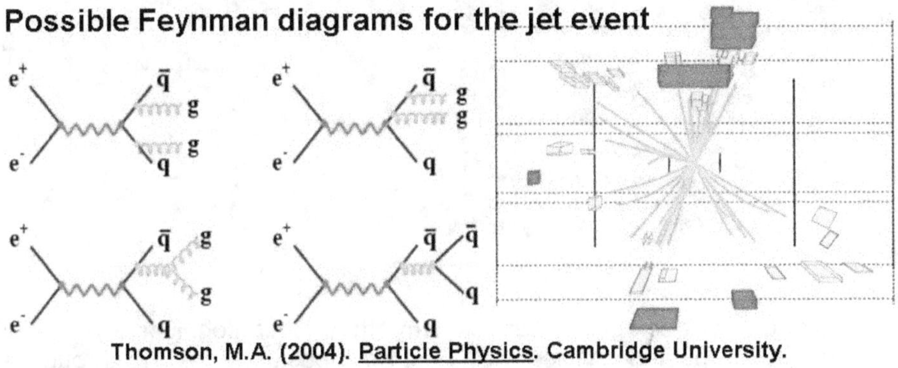

Possible Feynman diagrams for the jet event
Thomson, M.A. (2004). Particle Physics. Cambridge University.

e-Lepton Anagenesis
(e+)
(e−)
Photon Meiosis
Collision Energy
Gluon
Divergence
u−⅔
u+⅔
bg
yg

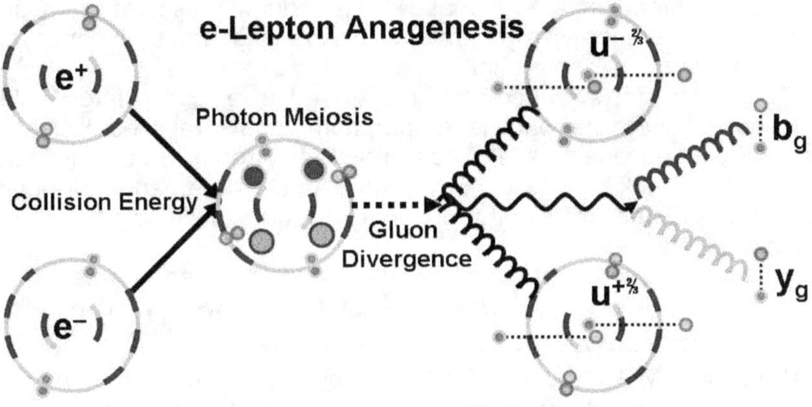

8.5: e to q + g Anangenetic Diagrams

---

31 Breinig, M. & Hitchcock, J. (2012). The Standard Model. electron6.phys.utk.edu/phys250/modules/module%206/standard_model.htm

32 Barnett, M. et al. (2014). Quark Confinement. particleadventure.org/quark_confinement.html.

Quantum Relativity

Thomson's image above is described as a four jet system.[33] The jets occur due to Pauli exclusion. We classify it as anagenic meiosis because it is linear evolution involving the creation of b:y photons that diverge/differentiate into –b(c:g) & +y(r:m) gluons.

In the vertical anagenesis diagram, in a charged e-lepton collision (pressure-volume), the energy appears as a photon (virtual being a squiggly line). The result is a quark-antiquark pair plus two more gluon jets. Those gluons are created by means of photon meiosis.

In confined spaces like a neutron, the jets can point into available host spaces. This creates transition particles as shown in down decay. This is the same decay that switches a neutron to a proton, and why we can't just ignore transition particles because they lack independent existence.

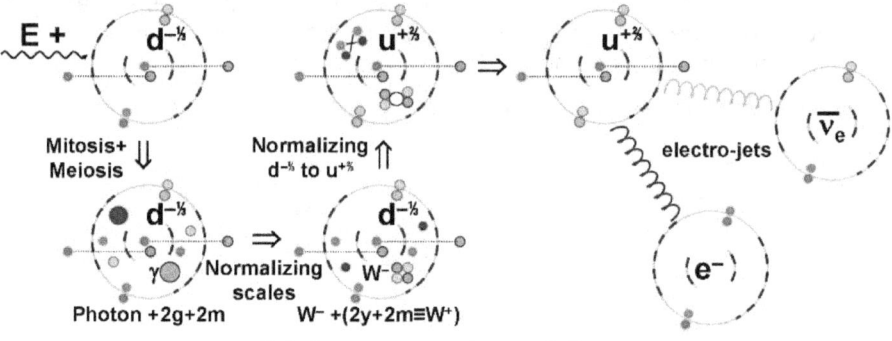

8.6: d to u + e + v Anagenetic Process

This gives topological reasons why they emit conditionally as an antineutrino or positron. When they emit as a positron, they go through a +W phase instead of a –W phase.[34] Procedural pressures of a space (volume) confine energy to generate form. Adapting form then conveys into working optimization and result.

Color roles get swapped from mu (μ)=gb to electro (e)=rg. Violating exclusion requires the particle to quantum tunnel (applied Heisenberg uncertainty) to escape as a jet. The proton to neutron process emits muons[35] due to down quark's b⁰ flavor. Feynman diagrams are handy for showing observations. These show why observations happen, giving the diagram predictive ability.

A Feynman version of this shows the quark confined in a hadron, and a common issue of which way to point the arrows. We can adapt a

---

[33] Thomson, M.A. (2004). <u>Particle Physics</u>. Cambridge University. hep.phy.cam.ac.uk/~thomson/particles/questions/Q16_answers.html.

[34] Irvine, J. (ret. Feb. 12, 2018). <u>Particle Physics Tutorial 11 - Particle Interactions</u>. antonine-education.co.uk/Pages/Physics_1/Particles/PP11/particles_ page_11.htm.

[35] Gorringe, T. & Fearing, H. (Jun. 18, 2002). <u>Induced pseudoscalar coupling of the proton weak interaction</u>. https://arxiv.org/pdf/nucl-th/0206039.pdf.

Feynman diagram-like form to show the enthalpy process (H=E+PV) in better detail.

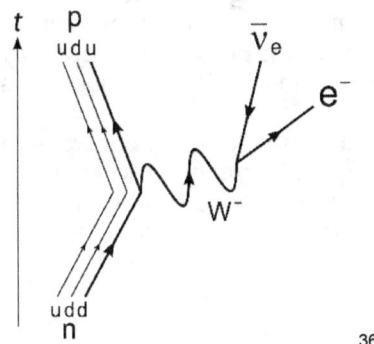

8.7: Feynmann Diagram of d to u + e + v

One thing the details in two dimensions do not easily show is that 2m+2y are asymmetric, interacting as synchronous like (m+y):(y+m). For the weak boson to be symmetric, the other must fail, and the fail most likely re-registers as a chiral pair defining a neutrino.

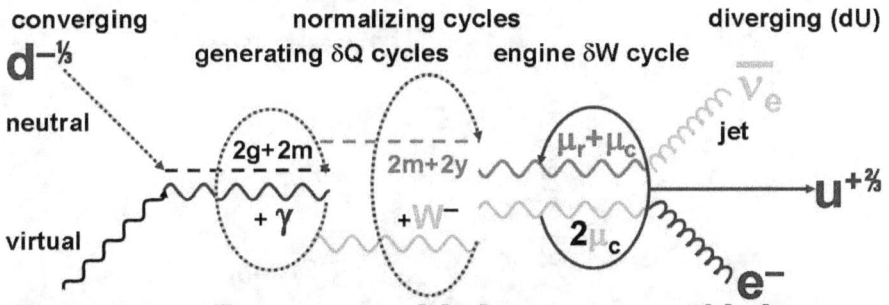

Energy + Pressure Volume = entHalpy

8.8: Enthalpy Process Diagram of d to u + e + v

Again, the jets occur because of quantum number violations (Pauli exclusion) and sticky color conditions. To escape they have to take the available b:y space they are strongly interacting with. Instead of quantum leaping to break this, they simply swap identities.

In transitional weak boson state, as with these muons, they are interchangeable with their electro-analogs (see pg. 114). That change results in spontaneous forced emission (jetting) of particle and energy. The new electro-identities stick. If we reverse the W-boson role, the y⁰

---

[36] Silverman, D. (Feb. 11, 2013). Weak Isospin and the Weakly Interacting Bosons (Force Particles). sites.uci.edu/energyobserver/2013/02/11/weak-isospin-and-the-weakly-interacting-bosons-force-particles/.

antineutrino (having –½ chiral isospin) becomes b^0 and the electron becomes a positron (y-bands).

Although not shown, the engine cycle loses a lot of energy in forging the Weyl fermions. That energy is contemporarily looked at as "fission." The diagram shows the fusion process is responsible for fission. From an atomic perspective it is fission because a nuclear isotope breaks down in the process.[37]

We tend to look at things in terms of our ways to dissect them. We see particles when we are smashing them into each other with intense added energy and pressure. Such violence is indeed commonplace, but so too is the passive version of these processes. The down decay diagrams are passive.

Interphase

In biology,[38] interphase is "the time between mitoses... During G1 (Gap 1), the cellular organelles and cytoplasm, including important proteins and other biomolecules, are duplicated. S (Synthesis) Phase is the point at which DNA is replicated. G2 (Gap 2) is spent double checking that no errors have been made during DNA replication."[39]

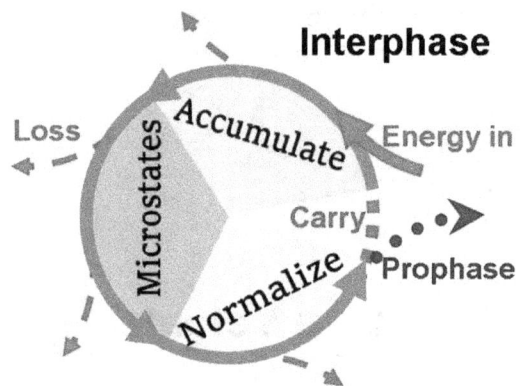

8.9: Interphase Process

---

[37] (ret. Feb. 12, 2018). Nuclear Fission. nuclear-power.net/nuclear-power/fission/.

[38] Riggio, G. (Jun. 2, 2017). 5 Stages of Mitosis. sciencing.com/5-stages-mitosis-13121.html.
Grimes, B., & Hallick, R. (Oct. 2004). The Cell Cycle & Mitosis Tutorial. http://www.biology.arizona.edu/cell_bio/tutorials/cell_cycle/cells3.html.
(2018). Mitsosis and Meiosis. Khan Academy. khanacademy.org/test-prep/mcat/cells/cellular-division/a/mitosis-and-meiosis.

[39] Baker, R. (May 18, 2017). Stages of the Cell Cycle - Mitosis (Interphase and Prophase). https://owlcation.com/stem/Stages-of-the-Cell-Cycle-Mitosis-Part-1-of-2.

Quantum Relativity

In physics, this sub-phase breakdown fits a phase (moment) energy cycle. Energy enters the system and accumulates with other excess energies. These energies pass through the extra-temporal microstates to metabolize. Metabolizing normalizes the information of intrinsic and transient energies. What hasn't escaped in the process compounds with future accumulation or triggers prophase.

Normalization is the observational perspective of smoothing into information equilibrium. This is computed using Schrödinger's equations,[40] which we later simplify into the Wheeler interaction (pg. 172 et seq.). Schrödinger's wave normalization function was designed to evaluate a wave function (ψ) to compute the probability of finding a particle along a trajectory axis (x) in time (t).

We are approaching from the opposite direction as if we are nature with all the answers. We aren't predicting. We are specifying by creating the wave cycle from the EMA-EMR information in the force energies. These smooth into the predictive normality.

$$\int_{-\infty}^{+\infty} \left| \psi \left(x, \int^c \left| \psi \left(t, \underbrace{\int_0^{2\pi} |\psi(\hat{k}, f_k + \Delta f_x)|^2 \, d\hat{k} = |\hat{k}_t^2| = +1}_{\text{cycle normalization = oscillation}} \right) \right|^2 dt = 1 \right) \right|^2 dx = 1$$

$$\overbrace{}^{\text{normalization of t in the wave over cycle } (\hat{k})}$$

8.10: Schrödinger Normalization of x in Wave (ψ) over time (t)

Schrödinger's probability density $|\psi(x,t)|^2$ is the observational frame in which force information differences sequentially oscillate and smooth relative to the duration (c) of time (t). Spacetime reference frame can be refined specifically because it is dependent on the central force oscillation.

Time resists microstate cycles and change of identity. Increments of time (δt) are defined by the period of cycle frequency (υ).[41] Cycle frequency increases directly to energy in the system per Planck's $E = h\upsilon$, so δt resistance decreases.

Energy adds into a system via relativistic momentum ($E^2 = E_k^2 + E_x^2$). The value origin is quantum force in kg m units. Applied to spacetime, energy becomes the time derivative $E = d\, cF/dt$. More practically F is observed as light frequency where information is the EMR-EMA detail. It is in this relationship where the particle wave confusion emerges. The correct

---

[40] Santra, S.B. (2013). Quantum Mechanics-Schrödinger Equation. iitg.ernet .in/santra/course_files/ph101/QM-02.pdf.
Fitzpatrick, R. (2010). Normalization of the Wavefunction. farside.ph.utexas .edu/teaching/qmech/Quantum/node34.html.
Nave, C.R. (2017). Particle in a Box. Georgia State University. hyperphysics .phy-astr.gsu.edu/hbase/quantum/pbox.html.
[41] OpenStax College (2018). College Physics. Rice University. courses .lumenlearning.com/physics/chapter/16-2-period-and-frequency-in-oscillations/.

interpretation of the relationship between frequency and relativistic momentum is temporal dilation.[42]

In each increment of time, energy can be added or removed from the system. In each increment of time, intrinsic and transient information goes through the microstate sequence and to a degree smoothes into one complementary sequence of information.

Generally speaking, the transient energy (f_x) is significantly less than the intrinsic energy (f_k). They proportionally mirror each other. Proportional mirroring gives the transient energy the identity of the interaction—the bands defined by attributed value like absorption bands from a reflection.[43] Even if the energy passes straight through, it passes during a microstate cycle, influences and is influenced.

Excessive application of force, as with acceleration, causes an excess in oscillation attempting to normalize the information. This and pressure preventing escape, lead to identity loss or transformation (k_t→anagenesis). If we could have all the information, we could also identify exactly which identity oscillation applies at specific points in the timeline.

Eigenstates

Lower case psi (ψ) represents the complete wave function of an identity—the Schrödinger equation. The function is the cycle of all energy distribution sets (microstates). Each set consists of all the color change functions acting as containers defining the whole identity. Each change container is a distinct wave function commonly called an eigenstate (Ψ).[44]

Each eigenstate in each set has a scalar energy value ($\hat{E}$). This scalar acts like a Laplacian in that it contains the EMA-EMR information: $\hat{E}=\hat{A}+\hat{R}$. The sequential sum of these for the set is the Hamiltonian operator ($\hat{H}$).

Ordinarily we think of the scalar just giving the wave function magnitude. What we actually have is a feedback system. The wave function has a smoothed operational perspective. The energy information is anomaly applied to the wave function. If the anomaly is too extreme, it changes the wave function of identity.

$$\hat{H} = V(r,t) - \frac{\hbar^2 \nabla^2}{2m}$$

---

[42] Acosta, D. (2006). Relativity 4: Relativistic Momentum. University of Florida. http://www.phys.ufl.edu/~acosta/phy2061/lectures/Relativity4.pdf.
[43] Blair, W.P. (ret. Feb. 15, 2018). The Basics of Light. Johns Hopkins University. http://blair.pha.jhu.edu/spectroscopy/basics.html.
[44] Simons, B. (2009). Operator methods in quantum mechanics. U. of Cambridge. tcm.phy.cam.ac.uk/~bds10/aqp/lec3_compressed.pdf.
Simons, B. (2009). Handout Chapter 3. U. of Cambridge. tcm.phy.cam.ac.uk/~bds10/aqp/handout_operator.pdf.

Quantum Relativity

Traditionally, $\hat{H}$ is broken down as the sum of kinetic and potential energies. Potential appears as a vector radial position in time=V(r,t). This potential is the effect of extrinsic forces acting on position and motion.

Kinetic energy appears as the absorption/intrinsic space defined by the Laplacian operator in a mass. The negative here resolves as generic i^2, the change function diverging container value (m) into the Laplacian. This is like giving a bag value and shape by putting things in it.

Technically, all virtual particles have complex (imaginary) mass functions which are attributed by context. Weak bosons normalize by equivalence into ordinary mass. Through the conduct of eigenstate evaluations, the complex forms must be explored. This is especially true for computing confined mass from the quantum details.

That imaginary element affects the divergence pattern of i^2. The divergence can be into a surface in or out of phase, volume, or a combination of these. Specific surface-volume combinations satisfy the GFE in the degree to which that eigenstate applies to the total wave function. Just as we overlay these maps in degrees to illustrate the fields, those same degrees affect the emerging mass value.

9. Relative Field Theory

Wheeler explained Einstein's gravity succinctly: "Spacetime tells matter how to move; matter tells spacetime how to curve."[1] Change functions are intrinsic to matter and define exactly how spacetime is shaped. Among these, the most fundamental is gravity: spacetime contracting into a linear pressure-order vector.

The second law of Thermodynamics requires disorder to establish order. The simpler your approach in one direction, the more complex it will be in others. This is why Thermodynamics applies broadly across Phase Mode. Its principles infest everything from greatest simplicity at the Möbiverse to extreme diversity of Use Mode.

Gravity is an independent concept from which mass emerges as a dependent variable. Its change function rotates it by spinor into a complex plane with the angular axis. The angular axis is fixed by being derived from a real quantum number (π in μ_0). The radian angle between these axes (φ) is the spinor unit setting the linear constants to local context. This is a tad more complex version of Einstein's space curving in idea.

The nature of gravity is a direct function of how matter is constructed. Like any other interaction, gravity is not just one thing: it is a whole class. As a class dependent on a hierarchy of spacetime constructs, it is emerges repeatedly in the hierarchy of vectors. It is first, fundamental, emergent, and ultimately last.

Quantum Gravity

In the sequence of matter, gravity as perfect enfolding order (singularity) is the first and ultimate field. We are not just trying to explain gravity in quantum terms.[2] Singularity of any size is a quantum number. For this reason, anything that can quantize into a fully occupied space (singularity) is fundamental and renormalized by its horizon.

The first instance of linear contraction establishing order is a primordial enfolding brane—a singularity surface tension with no depth. We will call this brane gravity, and its field equation the BFE. Einstein used the BFE to give value as stress energy curving spacetime into gravity.[3] Einstein makes regular use of branes as action containers ($/m^2$).

---

[1] Wheeler, J.A. & Ford, K. (1998). Geons, Black Holes, and Quantum Foam: A Life in Physics. NY: W.W. Norton & Company, Inc.
[2] Rovelli, C. (2008). Quantum Gravity. scholarpedia.org/article/Quantum_gravity.
[3] Johnston, W.R. (Nov. 3, 2008). Calculations on space-time curvature within the Earth and Sun. www.johnstonsarchive.net/relativity/stcurve.pdf.

$$8\pi G T_{\mu\nu}/c^4 \quad \text{BRANE = surface manifold} \Rightarrow \frac{2G\,\mathcal{E}_A}{c^4} = \left(\frac{2G}{c^4}\right)\left(\frac{4\pi}{\mu_0}\right) \quad \text{Boundary}$$

9.1: Einstein's Brane Field Equation

The BFE is derivative of the Poisson-Gauss field equations (see pg. 20). Gauss's field ($g_t = /s^2$) is a temporal container dilating a Laplacian distribution (∇=meters). $\nabla \cdot g_t = 4\pi G\,\rho_m$ is the linear acceleration of gravity valued by mass density (ρ_m=mass/V) applied to linear permeation (G).[4]

Renormalizing

Renormalizing is the process of limiting an infinite into a finite frame or group.[5] Here, renormalizing is confining (losing the original identities of) potentially infinite quantum complications into relativistic unit frames (quanta). These units are simply additive and/or subtractive. It is like finding a common denominator.

Strong bonds forming flavor volume and entanglement band pairs are forms of strong renormalization as their unit scales never change, but their arrangements do. The most obvious renormalizations are weak charge, helicity, and hadronization of mass. Interactions of those classes thereafter are simply additive/subtractive.

Gravity cannot renormalize[6] unless you count quantizing in a change proportion to form a color charge. Other than that, linear and angular/EM forces are subject to permittivity and permeability. Each is a quantum number subject to Pauli exclusion. The ideal state fills one and leaves the other attributed in the perfect order of singularity.

Every field is subject to permittivity/permeability. Every field can be condensed, dilated, or filled up to full occupancy. Interactions renormalize, but their fields do not. Their fields instead can quantize to form new matter or go the distance to singularity.

---

[4] Carroll, S.M. (Dec. 1997). Lecture Notes on General Relativity. UC Santa Barbara. https://arxiv.org/pdf/gr-qc/9712019.pdf.
[5] Baez, J. (Dec. 9, 2009). Renormalization Made Easy. math.ucr.edu/home/baez/renormalization.html
[6] Klauder, J.R. (Feb. 1975) "On the meaning of a non-renormalizable theory of gravitation." General Relativity and Gravitation. Vol. 6, Issue 1, pp 13–19. Springer.

Poisson used scalar potential (ϕ=/ms^2), setting g_t= $-\nabla \cdot \phi$. This gave $\nabla^2 \cdot \phi = 4\pi G\, \rho_m$, a complete Laplacian to distribute the acceleration of gravity.[7] GFE manifolds are distribution containers, making space the active component.

A straight-forward m^2 Laplacian is a distribution in 3-D space easily defining a spherical volume. The differential distribution is a temporally normalized change function containing the space.[8]

Gauss and Poisson appear to have seen time as the active agent: dilation acting on space. Einstein took the perspective of space being the container, its curvature the action.[9] Change shapes space into action on other spaces.

$M_{mn} = i(x_m \partial_n - x_n \partial_m)$ is a Laplacian generator symmetry.[10] It coincides with our change function chirality (imperfect mirroring), such that $\mu\nu$ is chiral of $\nu\mu$. For us, $\mu\nu$ = disorder:order = cyan. The difference affects the shape of local fields and details of strong interaction.

A brane has in and out of phase forms—seemingly a topological paradox. The out of phase form is imaginary until put into a more complex interaction. In phase, all points on the surface are the origin acted on equally from all directions.

This is like the elasticity of a balloon surface and why the actual horizon is limited by the enfolding manifold value (pg. 15). Just as a balloon without filling collapses into itself, a singularity of any magnitude enfolds out of existence without interaction.

A singularity of any magnitude is order annihilating itself. At a primordial level the only thing it can annihilate is itself.

$$\upsilon_\gamma = \frac{1-\eta}{h} \left(\mu_{n-1} = \frac{m_{n-1}^2}{\nu_n} \right) \left(\frac{\nabla^2}{t^2} = c^2 \right)$$

mu at least = lesser m^2 in Laplacian spacetime units

upper magnitude singularity

thermodynamic efficiency $\quad \eta = \dfrac{\delta W = PdV}{\delta Q = TdS} \quad$ Ejected value / Enfolded value

9.2: Value Redistribution Function

Redistribution feeds the energy back from the bubble's actual horizon toward the singularity at the speed of light. This by itself is too slow to

---

[7] Brown, K. (2017). Poisson's Equation and the Universe. mathpages.com/home/ kmath711/kmath711.htm.
[8] Grinfeld, P. (Feb. 12, 2017). What is the Laplacian? Philadelphia, PA: Drexel University & Lemma. youtube.com/watch?v=4J74tquQ7jU.
[9] Overduin, J. (Nov. 2007). Einstein's Spacetime. einstein.stanford.edu/SPACETIME/spacetime2.html.
[10] Di Francesco, P.; Mathieu, P; & Sénéchal, D. (1997). Conformal field theory. Graduate texts in contemporary physics. Springer.

maintain the singularity even if it could interact. Bubble interactions also modify locations of singularity and current distribution. Without local disorder to maintain a singularity, the interactions rapidly steal all that value.

Singularities with their horizons have local and long-ranged input-output paths. Local input must be mass-oriented, providing disorder to prevent enfolding without limit to annihilation. The input can only increase enfolding if it can occupy the available permeability space and be subject to the information changes to S=0 needed. Primordials cannot do this.

Unlike a black hole, a primordial cannot "feed" off of other matter, and neither can feed of light. Primordials cannot grow, but they must be maintained. Unfolding a primordial by our standards is pretty local, but inconsequential unless it is outside of interaction. It emits as a longitudinal (gravitational) wave photon.

A collision style gravitational wave is like a splash in the fabric of spacetime. LIGO (Laser Interferometer Gravitational-Wave Observatory) uses a split laser beam to observe differences in otherwise equal lengths at right angles to each other.[11] These are not easy to observe at any level. A primordial gravitational wave follows a reflective path. It starts at one horizon, passes through and flips at its origin as if it passed through a lens before propagating away entirely.

The local mu-interactive cycle sustaining a singularity is an electromagnetic function. For a singularity to form and continue its existence at any magnitude, it must maintain enough interaction to maintain this electromagnetic field.

Einstein saw the enfolding, and it is possible with his static universe he awkwardly got the idea of its unfolding. Unable to resolve this seeming paradox, his gravitational singularity enfolds impossibly to infinite.[12] No doubt what we are describing here contributed to his thinking curvature is causal. Depends which gravity.

GFE Fields

Einstein's geodesic field equation (GFE) consists of interacting rectilinear surfaces.[13] If we were to migrate 4π from the BFE curvature to the GFE side, these surfaces become spherical. Rectilinear (Euclidean) is a quality of the Laplacian.[14] Each of the GFE manifolds is a Laplacian

---

[11] (ret. Apr. 3, 2018). What is an Interferometer? https://www.ligo.caltech.edu/page/what-is-interferometer.
[12] Claes Uggla (Oct. 17, 2017). Spacetime Singularities. einstein-online.info/ spotlights/singularities/index.html@searchterm=None.html
[13] Hitchin, N.J. (1974) Compact Four-Dimenstional Einstein Manifolds. projecteuclid .org/download/pdf_1/euclid.jdg/1214432419.
[14] Holloway, R. (Nov. 7, 1999). The Laplacian in different coordinate systems.
personal.rhul.ac.uk/uhap/027/PH2130/PH2130_files/laplacian.pdf.

generator, which confuses things. Here, **surface gravity** emerges from the interaction of two surfaces that confine (conceal) their origins.

The problem with Einstein's GFE is common in physics: ambiguity. It is incredibly easy to get lost in so many ambiguities. One of the ambiguities is the issue of volume. By acknowledging the brane quality of these manifolds, we do not exclude the volume, we just confine it. The content of the volume becomes irrelevant.

By confining the volume, we now see the GFE consists of interacting surfaces (/m²)—a Riemannian is "made up of an infinite of Euclidean spaces."[15] The linear values derived from these surfaces are radii of order (o) and disorder (a= attributed). The geodesic function is gravity attempting to quantize into the perfect S=0 order of singularity.

Quantizing requires differentiating EMR as + (contracting) and EMA as − (expanding) values into the ± distribution (confinement) of j -entropy. It also requires either equalizing them or satisfying another boundary condition. Equalizing them zeroes the GFE out. What remains is the enfolding effect of $r\sqrt{2}$ displacement. The GFE manifolds can be made semi-conventional to illustrate this.

Without forgetting their origins, we can reduce each segment of Einstein's GFE to a radius by taking its numeric square root as shown below. Those radii translate perfectly into ordinary spherical topology to show what part is confined, what part is loose with directional surface, what exactly is being displaced, microgravity and continuity into generalized gravity with no upper boundary.

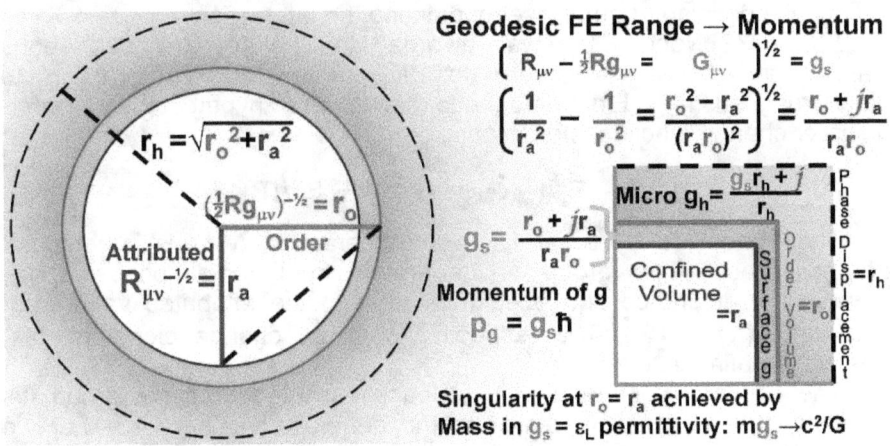

9.3: GFE Boundaries and Conversion to Momentum

This breakdown is a lot easier to convert into other applications like momentum. We can relate this to distribution ($g_S^{-1}=\nabla$), from there to Gauss and the acceleration of gravity: $g_t/g_S = 4\pi G\, \rho_m$ (noting dilation=g_t/s² and

---

[15] Besse, R.L. (2000). <u>Einstein Manifolds</u>. Berlin, Germany: Springer-Verlag.

mass density=ρ_m). For the sake of radial symbolic consistency we can say $r_k \equiv g_s^{-1}$ to reflect influence of confined change variables (k).

It also links directly with linear permittivity (a Schwarzschild $\varepsilon_L = c^2/G$). Permittivity is typically used in electrodynamics to indicate "the capability of the vacuum to permit electric field lines."[16] Generically, permittivity is the capacity of space to receive value. It shows we can fill this space with mass (m • g_s) to permittivity.

There are two immediately obvious paths to singularity: mass accumulation and equality of ordered and attributed values. The energy of momentum ($E = g_s \hbar c \rightarrow mc^2 \equiv h\upsilon$) opens the GFE to the full range of applications and permittivity quantizations. Hiding in the midst of this are two more major fields, only one of which is properly a function of gravity.

There are two degrees of microgravity. The first degree is a localized effect of the electroweak field. The second is the continuation of that as an open-ended long-wave. Microgravity and the Fermi surface are tricky and require us to resolve the electroweak field first.

The Electroweak Field

The weak interaction is a set of conditions enabling flavor change (charge-parity/CP violation). It derives from Fermi's theory for contact particle exchange responsible for radioactive decay and subsequent atomic fission.[17] The phenomenon divides into mediation particles and the field space defining or otherwise modulating the interaction.

The 1979 Nobel Prize was awarded to Sheldon Glashow, Abdus Salam, and Steven Weinberg for linking the weak interaction to electromagnetism.[18] Electromagnetic fields are hypercomplex—order-disorder change function interactions.

$$w_{kh} = i(r_k \partial_h - r_h \partial_k) = s^2/m^2$$

The spacetime potential for weak interaction (w_{kh}) is a temporal Lorentz group. The group is generated from radial phase space ($r_h^2 = r_A^2 + r_o^2$) difference squeezed (dilated) into the attributed value sub-spacetime ($r_k = g_s^{-1} = \nabla$). The respective confined change elements (h,k) interact to define time (s^2).

The confined volume of (r_A) is space excluded from the group. Its confined change function relates to the general container ($j_A = /j_O$) we expect. Temporal emergence is seen here in the renormalization of weak charge.

---

[16] Serway, R.A. (2017). College Physics. Vol. 2. 9 ed. Content Technologies.
[17] Wilson, Fred L. (December 1968). "Fermi's Theory of Beta Decay". American Journal of Physics. 36 (12): 1150–1160.
[18] The Nobel Prize in Physics 1979. Nobel Media AB 2014. nobelprize.org/nobel_prizes/physics/laureates/1979/.

Quantum Relativity

The weak charge sign is set by $i/j_O = ij_A = \pm \rightarrow s^2$ otherwise washing out in the equivalent s^2. Temporal abstraction isolates the charge unit to relative scale, like we used with microstates. Weak bosons and leptons contain the same degree of scale (e.g. parts and interactions). They set the normalized unit reference.

Charge from this point simply adds and subtracts—just as strong interactions affect mass up to hadrons, then they simply accumulate additively. A Weyl fermion as a strong volume is a magnitude greater than gluons and photons. None of these has a weak charge by itself though. Put into a weak context, Weyl fermions get ½ where gluons and photons get ⅙ charge equivalences.

This charge is temporal (renormalized). It sets the aspect perspective. It doesn't redefine the band identities, but does affect orientation of matter created in those bands. In electroweak theory this is the "weak mixing angle" applied to Cabibbo's matrix to compute spontaneous symmetry breaking.[19] This accounts for complex angles flat diagrams cannot show, like why an antineutrino (chiral isospin) occurs in down decay.

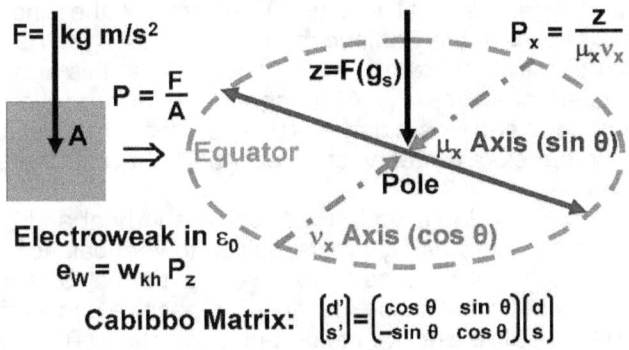

9.4: *Electroweak Field Explains Cabibbo Matrix*

The density of this space is its "force." That density contains intrinsic ($v_x = jmG/c$) and transient qualities (e.g. heat). The intrinsic value can be described as linear (x) mass value. The field is an elevation-related horizontal mass density consistent with specific weight.

Like heat, this is of a lesser magnitude than the force of gravity. Heat density is also fairly simple mathematically ($\mu_x = iTkc/\hbar$). We find using standard kg m s units best for purposes of compatibility and reducing confusion. Boltzmann's constant (k=1.38064852 x 10^{-23}-23 kg m^2/s^{2}K)[20] converts heat in Kelvin to kg m s units, with a consequence. Again, constants are not just numbers. They actually do things.

---

[19] Cvavb, C.R. (June 2000). Is the Cabibbo Angle a Function of the Weinberg Mixing Parameter? International Journal of Theoretical Physics, Vol. 39, Issue 6.
[20] (ret. Feb. 22, 2018). Boltzmann Constant: k. https://physics.nist.gov/cgi-bin/cuu/Value?k.

The imaginary elements are added to show time emerging in $ij = s^2$. This is particularly significant to Fermi surfaces and the effects of supercooling on friction, conductance, etc. Reducing heat also reduces the temporal resistivity of this space. It becomes less resistant to i-entropy (e.g. disorder and electrical current).

The imaginary elements are also relevant to the fact that you can't just convert everything willy-nilly to basic SI units (kg m s). There are consequences, you just need to understand which axes rotate and how. These complex forms are made real by confining context.

The densities focus more intensely than gravity alone. The strength of the weak interaction is 10^{-6} that of the ordinary strong interaction.[21] Coupling constants are used to describe the strength of interactions. A coupling constant of g<1 is considered weak. Fermi's is computed using muon life and W-boson mass: $G(F)/(\hbar c)^3 = 1.1663787 \times 10^{-5} /GeV^2$.[22]

The electroweak spacetime ($w_{kh} = s^2/m^2$) is a directional density potential. Populated with substance, the substance interacts with itself as an area $A = \mu_x v_x$ defined by heat (μ_x) and order (v_x) axes. While circular, we tend to look at these as rectilinear (x). They provide the sine and cosine quadrant applications in the Cabibbo matrix.

The force of gravity (shown as the z-axis) put into this area defines the actualized system pressure (P_x). Applying this pressure to the electroweak spacetime potential provides a degree (U≤1) of electromagnetic permittivity ($U\varepsilon_0 = w_{kh} P_x$) as the electroweak vector field value. At U=1, the electroweak field quantizes.

The electroweak field fluctuates with change in heat and content. The charge sign is the native identity predisposition toward balance: + @ $v_x > \mu_x$, 0 @ $v_x = \mu_x$, − @ $v_x < \mu_x$. Ideally you want a perfect balance that can manage imbalance without losing identity. Too positive and it enfolds out of existence. Too negative and it unfolds out of existent. Too neutral and it triggers either way with little provocation.

This is essentially why the universe is dominated by atoms. Atoms keep their electron disorder fields on the outside held in tight conformity by the weak charge interaction with a proton. Energy changes affect how tightly packed these relationships are. New electron orbit groups compact existing orbit groups, and new members to a group enlarge the group size.

MicroG & Fermi Surface

When the group can't expand or contract due to structure and pressure conditions, and especially at boundaries, new qualities become evident. One of these is the creation of a space called a Fermi surface—a

---

[21] Nave, C.R. (2017). Fundamental Force. Georgia State University. hyperphysics.phy-astr.gsu.edu/hbase/Forces/funfor.html.
[22] (June 2015). "Fermi coupling constant." The NIST Reference on Constants, Units, and Uncertainty. physics.nist.gov/cgi-bin/cuu/Value?gf.

quantum-free (available) weak field remnant. The most notable Fermi surface applications in material science are viscosity, phase states (solid, liquid, gas) and resistivity in electrodynamics.[23]

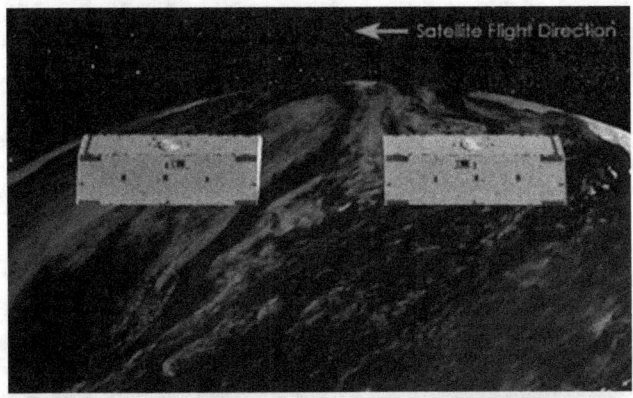

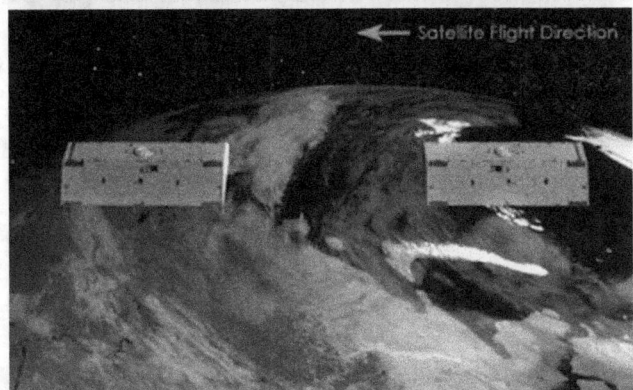

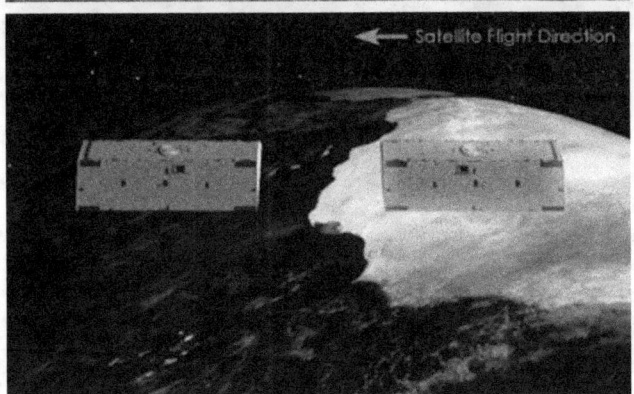

9.5: GRACE Satellites Ride g-Waves

---

[23] Dugdale, S.B. (Apr. 18, 2016). Life on the edge: a beginner's guide to the Fermi surface. The Royal Swedish Academy of Sciences. iopscience.iop.org/article/10 .1088/0031-8949/91/5/053009/pdf.

Quantum Relativity

The electroweak field extends beyond the region of surface gravity. Beyond the range of surface gravity, z in the electroweak field is defined by phase displacement. This is the generalized microgravity by which the rest of the universe interacts with us. Micro is misleading as it is long wave.

Microgravity appears in the EFE as $G_{\mu\nu} - \Lambda g_{\mu\nu}$. It is the relativistic displacement ($G_{\mu\nu}$) minus the environmental phase density ($\Lambda g_{\mu\nu}$). The displacement makes it quasi-directional. Quasi-direction acts to put objects into contextual aspect. Between this and anomaly causing horizontal interaction, near orbits easily become trajectories.

Microgravity generally acts horizontally on objects in this space. It acts on approach by or recession of other charge spaces including attempts to escape or re-entry objects. For a singularity, it defines how things can enter or be discharged from the singularity's spaces, but is color not weak charged. It can set the input/output channels quite far away from the actual object.

GRACE is a pair of NASA satellites connected to each other (see above). They orbit at 220 km (137 miles). The distance between them is a measure of horizontal field differences resulting from density variations. The lead satellite is slowed by increased density causing the distance to contract, and then accelerates by decreased density causing distance to expand. [24] Something like roller coaster cars adjusting to the terrain.

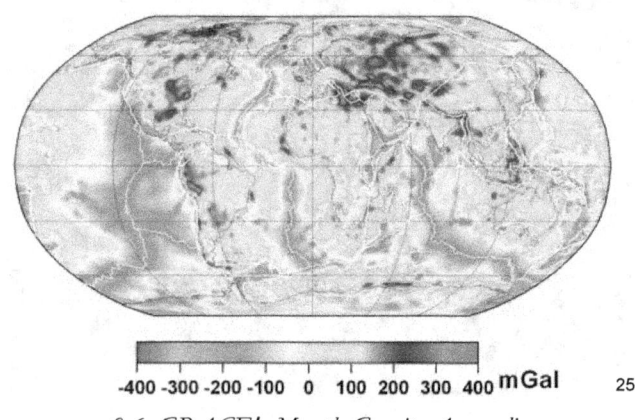

9.6: GRACE's Mantle Gravity Anomalies

They are accurately observing horizontal fluctuations of substantive order (v_x) and heat disorder (μ_x). The fluctuation is mapped as the anomaly of gravity. The pressure of the electroweak field affects the rate of local surface gravity intensity—specific weight. The surface of the mantle is the

---

[24] Ward, A. (Mar. 30, 2004). <u>Gravity Recovery and Climate Experience (GRACE)</u>: "The Workings of GRACE." earthobservatory.nasa.gov/Features/GRACE/.

[25] Reigber, C. (Jan 20, 2004). <u>First GFZ GRACE gravity field model EIGEN-GRACE01S released on July 25, 2003</u>. op.gfz-potsdam.de/grace/results/grav/g001_eigen-grace01s.html.

confinement surface. Take away everything else and surface gravity is constant.

The anomaly of gravity shows changes in the ordered brane from which disorder is subtracted to define surface gravity. This subtlety is like being subjected to an error function while in orbit. Generalized gravity from phase displacement has no direction without interaction. The region of microgravity provides degrees of incidental interaction. You can orbit in this region, but the anomaly will act as friction eventually causing orbital decay.

Degeneracy

I was at first almost frightened when I saw such mathematical force made to bear upon the subject, and then wondered to see that the subject stood it so well.

—Michael Farraday, Mar. 25, 1857
letter to James Clerk Maxwell
re: "On Faraday's Lines of Force"[1]

---

[1] (1884). <u>The Life of James Clerk Maxwell: With Selections from His Correspondence</u>. Campbell, L. & Garnett, W eds. p. 200. archive.org/details/lifeofjamesclerk00camprich.

10. Confined Morphology

Virtual matter provides focal values whose interactions construct confined spacetimes through the process of hadronization. Once matter is functionally confined, it becomes subject to a new range of topological constructions and interactions.

In this chapter we will examine the root of structure formation, the resulting topologies, and interactions defining the complex field behaviors of confined matter. This includes such topics as nuclear structure, magnetic induction, electron orbits and the limits of atomic scale. We then advance into the quantized momentum and vector systems that shape the cosmos.

Trionic Bonds

A lattice cell is the interaction and arrangement of discrete points into a regular geometry.[1] A nuclide is the set of protons and neutrons defining a nuclear isotope.[2] A nuclide lattice is thus the specific geometric arrangement of protons and neutrons. The rules of arrangement show emergent physical properties like magnetic induction, susceptibilities, and isotope limitations.

Nuclear structure is typically depicted as a random mix of nucleons in a ball shape. The assumption is that nucleon interactions are quantum, so it doesn't matter. Nearly everything quantum in nucleons has renormalized. Spacetime specifically has been renormalized such that further interactions are no longer defining space from scratch, but rather defining points in space.

The only quantum components not entirely confined are the trionic band edges and surfaces used to strongly interact with other nucleons. These bands and resulting strong interactions (type VI see pg. 117) are subject to uncertainty including microstates and color changes. Those uncertainties enable renormalized qualities like weak charge, but do not violate them.

The rules of type VI strong interaction are regulated by the color configurations of trion geometry. For this reason, we call them trionic bonds. The original strong bond formed flavor volumes and could occur in pairs (rg|cm, gb|my) or triplets (rgb|cmy).

---

[1] Gross, R. & Marx, A. (2014). Solid State Physics (Festkörperphysik). Berlin: De Gruyter Studium. German language.
[2] Chieh, C. (Jan. 21, 2004). Nuclides. University of Waterloo. science.uwaterloo.ca/ ~cchieh/cact/nuctek/nuclide.html

Quantum Relativity

Through the stages of development, triplets became the stable preference for space-sharing strong bonds. Confinement has also created an unusual band condition. These define edges and between the edges is an anti-band surface.

If the bands are colors, the surface is the subtractive combination of neighboring colors. For example, between r & g bands is a y surface. Leptons also have surfaces like this, but their context renormalizes that surface into temporal.

The trion surfaces are attributed spaces. As a consequence, the space of this interaction is both electroweak and strong. The electroweak role is generally restricted to charge, though energy changes can evolve that to nuclear reactions affecting trion bonds. Each trion bond is actually a double bond between two bands with opposite edges and their complementary band surfaces.

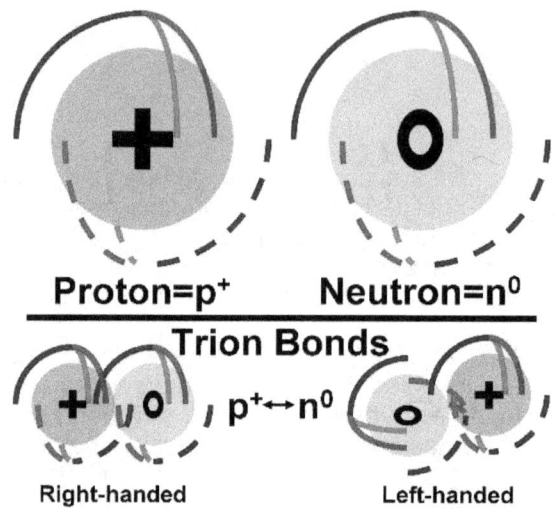

10.1: Trionic Band Edge-Surfaces and Bonds

Right-handed bonds join the in a common j-spin direction. The left-handed bond causes the sinusoidal s-shape of disorder (i). Left-handed reveals how these bands function.

In the sequence of microstates, bands become real while the main identity parts become null. Without this actualization of value, there is no interaction. With actualization of band values, the left-handed bond risks breaking. Energy change affects frequency, and a complex oscillation can have significant consequence.

Each trionic bond uses up one band and one surface for each member's trionic geometry. That geometry allows for only three bonds of this nature. As a renormalized interaction, the trionic bond is also subject to ordinary energy distribution.

In microstate evaluations we smoothed quantum numbers by rounding e up to 3 (pg. 80 et seq.). Here we can't round due to renormalization. It is

Quantum Relativity

no longer just a quantum number to smooth. It is an ordinary real value we cannot ignore because it is defining a boundary condition. Without this boundary, nuclides would not need structure.

Structure Groups

Geometric structure is a classical concept we can all relate to easily. Structures are vital to functionality. Nuclide structures are classical solid geometry concepts that are held together by quantum elements confined in the process (renormalized). A regular geometry optimizes the energy use of the parts but can affect other functions.

In engineering, spheres are ideal structures followed by triangles and squares.[3] Regular/Platonic solids are generally considered to be Euler's with 4, 6, 8, 12, and 20 faces. The sphere is also a regular solid, except it represents the quantum concept of a point-surface. In mathematics, points are generally brushed aisde by assigning infinite points to a line, then line/edges enclose surfaces that enclose volumes.

Pentagons aren't remarkably sturdy, but in an icosahedron they form an approximation of a sphere. This is the one regular solid you cannot connect with a continuous non-crossing line. If there were such a thing as a g-orbit for an electron, it would attempt to follow this shape and fail. The dodecahedron is a near approximation of a sphere with triangles.

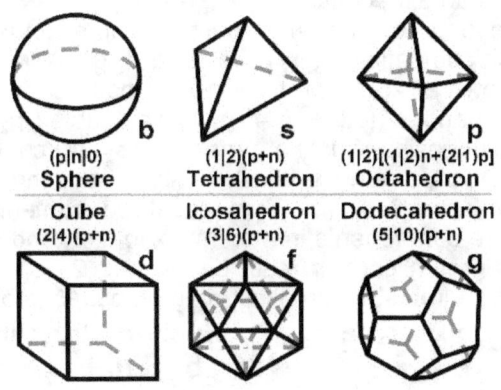

10.2: *Regular Solids Give Structural Rules*

These are in order of vertices, which roughly coincides with ideal structure sequence. Hyper-hedrons are a variation of these. Hyper-

---

[3] Chinnis, D. et al. (2013). Exploring Structural Engineering Fundamentals. teach engineering.org/content/cub_/lessons/cub_trusses/cub_trusses_lesson01_presentation_v5_tedl_dwc.pdf.

Fyon, A. & Ginn, E. (1999). Professor Beaker's Everyday Structures and Systems. steps.ie/StepsToEngineering/media/StepsToEngineering-medialibrary/Volunteers/Activities/Everyday-structures-and-systems.pdf.

hedrons are connected internally like a building architecture. Baez provides hyper-hedron[4] illustrations that are very stable because they build up with triangular structures.

Nuclides can certainly structure themselves this way too, but at significant cost. It would dramatically affect the emergent fields and create other complications, but a worthy pursuit to explore. It would also be more consistent with pure quantum interactions than this level of renormalization.

The trion suggests hyper-hedrons are the exception rather than the rule. Trions are more conducive to surface vertices and edges of polyhedrons forming into layers. Thanks to microstate energy exchanges, they don't need to provide all the edges simultaneously.

Due to renormalizing in structure, energy can be treated as distributed in equilibrium across the edges as if they were simultaneous. Things get a little trickier when the nucleon structures form layers of nuclide shells. Each solid describes a symmetric or asymmetric nuclide. The asymmetric nuclide would consist of half the points of its symmetric cousin, or with hyper-hedron like connectivity.

Ordinarily we may assume that an asymmetric must be set upon a symmetric and that they would occur in the order of sequence stability. They definitely occur in order of sequence stability, but not necessarily by symmetry or handedness. For example, a lone or randomly injected left-handed would be considered unstable. As part of a nuclide core, however, left-handedness is an expectation.

Each of these is shown with possible architecture contents and assigned an orbit-compatible sign (b, s, p, d, f, g). A lone neutron in the core could be shown as $_0b_n$. A proton+neutron pair as $_1s_1$. A p-shell can be single right or left-handed (l=2n+1p→ $_1p_{1n}$) or a double ($_1p_2$).

Nuclide shells consist of unlike parts, so they will configure differently. A lone proton (H) is $_0b$; none explicitly is $_0b_0$. Assuming one neutron per proton, He fills $_0p_1$, and B fills $_0g_1$. The spatial requirements, however, limit the number of core components in a complex to 0, 1, and 2 protons—up to $_0p_1$. If there is a proton core, it must be H (1=$_0b$, 2=$_0s_1$, or 3=$_0p_{1n}$) or He (3=$_0p_1$ or 4=$_0s_2$). The iron structure illustrated on pg. 154 could be written:

$$\text{nuclide} = {}^{52}_{26}\text{Fe} \qquad \text{Nuclide Structure:} \quad _0b_0 + {}_1g_2 + {}_2f_2 + {}_3g_2$$

This structure is confined in an extreme mass density condition of $\rho_m = 3N_m/4\pi r^3$ (nuclide mass per spherical volume).[5] The spherical radian $1/2\pi \equiv s^2$, makes nuclear density a function of magnetic (μ_0) and linear (G) permeability in a change distribution ($s^2 = k^{-2}$ see pg. 172). The available

---

[4] Baez, J. (Nov. 12, 2006). *Platonic Solids in All Dimensions*. math.ucr.edu/home/ baez/platonic.html.
[5] Nave, R. (2017). *Nuclear Size and Density*. University of Georgia. http://hyperphysics.phy-astr.gsu.edu/hbase/Nuclear/nucuni.html#c4.
(2018). *The Atomic Nucleus*. cyberphysics.co.uk/topics/atomic/nucleus.htm.

space is a function of angular permeability ($U_A=f\mathcal{E}_A=4f\times10^{-7}$ kg/m s^2) in a distribution of the remaining change (/k^2):

$$\rho_m= [U_A/k^2] [\mu_0/2G=2.34378270633671\times10^{17} \text{ s}^4/\text{m}^2]$$

U_A/k^2= kg/m s^4 is the "emptiness" in quantum foam. The foam container is idealized by $\mu_0/2G$. The proper container is /2G. Magnetic charge projection is a significant part but is non-local, complicating everything. Only part of this non-localizing factor can participate in containing localized angular value (f$\mathcal{E}_A$). That is one challenge of realizing this idealization.

Foam is a relatable concept to understand how degenerate density works. It can be formed with trapped impurities adding to emptiness or removing from the container. The structure defines where what can go to manifest and trigger changes like neutron decay.

Presumably only part of U_A/k^2 is available. Satisfying the angular quantum number $\mathcal{E}_A$ may be localized, made ambiguous, and trigger only a partial discharge. Neutron star observations suggest a quantum number is satisfied where f$\mathcal{E}_A=1|e$ kg/m s^2 (π is already used by μ_0). Achieving 1 is the idealization fulfilled from a lesser ground state density, and e is the maximum for ground state densities numerically $\geq\mu_0/2G$.

This idealization and related functions can help establish G (see pg. 76). Diverse and contextual functions connect at idealizations like this by arriving at the same values. For example, we can use the Schwarzschild with an electroweak disorder of:

$$U=2\pi [w_{kh} = \ddot{u}(r_k\partial_h - r_h\partial_k)\to c^{-2}] = (c\hbar/\delta Q)^2 = /m^2$$
$$\rho_m = [2\pi w_{kh}] [c^2/2G]$$

Electroweak spacetime renormalizes into ordered nuclide density. This enfolds the structure with a distribution of disorder into a complex function.

QM Aspects

There are several approaches to nuclide structure. When one problem produces a class of working solutions, one must conclude the problem itself is a contextual complex to be analyzed in quantum terms. Elements of multiple approaches hold simultaneously true. Watkins lists several:[6]

Liquid/quantum drop model—a Gamow idea enhanced by Bohr-Wheeler to describe the nucleus as a smooth but fluid consistency (like a water drop). Their nucleus would be confined, contiguous, and non-compressible.[7]

---

[6] Watkins, T. (ret. Feb 25, 2018). Models of Nuclear Structure. San Jose State University. http://www.applet-magic.com/nuclearstruct.htm.

[7] Shupe, C. (2015). Nuclear models: The liquid drop model Fermi-Gas Model. atlas.physics.arizona.edu/~shupe/Indep_Studies_2015/Notes_Goethe_Univ/A2_Nuclear_Models_LiqDrop_FermiGas.pdf.

Shell models—follow electron orbit-like shapes or quantum renormalize away from those shapes.[8] Quantum is good for playing both.

Model of Independent Particles—treats the nucleus as an n-body problem, which fits certain quantum contexts.

Substructure/Cluster model—n-body problem contextually forms m-groups, again fitting quantum contexts.

Collective model—shape arises from collective motion, which is true except it is quantum motion so the shape is simultaneously defined.

Fluctuating Combinations model—fluctuation among all possible combinations.

Interacting Boson model—is among the most relevant and shell-related;[9] the electroweak spacetime component paths to magnetic susceptibility, induction, and degeneration.

Alpha Module model—Uses alpha particles (α=2p+2n) to compress/compact lattice/structure points. With heavier elements this is an extremely likely way to evolve the structuring to handle the massive numbers.

We omitted Lattice/Monte Carlo as a method of exhaustive computational experimentation. These all contain valid elements or concepts.

Quantum Dynamo

Nuclide structure is a quantum phenomenon renormalizing mass density. A lot of features are confined (hidden) in the process. Confinement also means we cannot think of our shell structures classically. They emulate classical mechanics, but in a quantum way—hence quantum mechanics.

As QM, we can go through all our conventional mechanical thinking and computations. We absolutely cannot forget that these measurements are applied to an imaginary axis in the system. This QM system can form a dynamo effect in degrees or perfectly (e.g. ferromagnetism).

The **dynamo effect** focuses angular momentum through the intrinsic charge field (μ_0) projection found in the degenerate density function ($\rho_m = [U_A/k^2]\,[\mu_0/2G]$). This allows energy to be kept in the system as a projected cycle that would locally violate density or other scale limitations.

The dynamo-effect is a geophysics theory in which spinning interactions of the core generate magnetic field lines. In contemporary thought, ferromagnetic elements with weak magnetic fields excite electrons

---

[8] Dayou, C. (Feb. 6, 2018). Statistical Model of Nuclide Shell Structure. NW University, China. medcraveonline.com/PAIJ/PAIJ-02-00050.pdf.

[9] Elliot, J.P. (1985). The interacting boson model of nuclear structure. iopscience.iop.org/article/10.1088/0034-4885/48/2/001.

Cejnar, P. & Jolie, J. (July 22, 2008). Quantum phase transitions in the interacting boson model. https://arxiv.org/pdf/0807.3467.pdf.

Quantum Relativity

in the convection currents of the mantle and core.[10] This is based on inducing electrical current by rotating a coil in and out of magnetic fields.[11]

| Shell A (10p⁺ 10n⁰) | Shell B (6p⁺ 6n⁰) | Shell C (10p⁺ 10n⁰) | Magnetic Field |
| Icosahedral | Dodecahedral | Icosahedral | Bands (4) |

SIDE 1

SIDE 2

Band colors correspond with colored stars in shell diagrams

10.3: Ferromagnetic Nuclear Shells to Field Lines

Electromagnetism covers a very large group of cross-phase interactions—namely interactions between order and disorder on quantum change levels. The result is bi-directional quasi-temporal fields consistent with h-entropy (phase).

Electrons and electrical current certainly play a role in energy conveyance and related fields. They are not the actual source of magnetism. The source is in the structural interactions of the degenerate nuclide. In part this can be viewed as extra-temporal transient weak charge interactions. While incomplete by itself, it gives a starting generalization.

The neutral charge in passes through the same electroweak space as the positive charge and conveys the expansion of both into a projected angular field. The conversion from angular to magnetic is simply a change in orientation relative to the origin (around becomes adjacent, recall constants on pg. 56). For this reason, Coulomb's angular and Maxwell's electromagnetic constants are inversely related: $4\pi\varepsilon_A = \varepsilon_0^{-1}$.

---

[10] Guzman, R. (Oct. 15, 1997). <u>Dynamo Effect</u>. astro.ufl.edu/~guzman/ast7939/glossary/dynamo_effect.html.
Bettex, M. (Mar. 25, 2010). <u>Explained: Dynamo Theory</u>. news.mit.edu/2010/explained-dynamo-0325.
[11] Tilgner, A. (Apr. 2, 2012). <u>The Twists and Turns of the Dynamo Effect</u>. https://physics.aps.org/articles/v5/40.

The linear expansion in time becomes angular. Angular force is object contained (m^3/kg). Neutralizing charge passing through that electroweak space swaps from object to volume oriented (kg/m^3).

The energy of magnetism is the energy in the electroweak spacetime. This includes intrinsic values from trion bonds, extra charge values,[12] and transient energies, like the force of electrical current in the solenoid of a bar magnet.[13]

Earth's magnetic field is most likely a product of weakly interacting (super) massive (degenerate) iron-like particles (WIMPs) excited by the active core and even more complex field conditions. Electromagnetism includes intrinsic and environmental influences.

How magnetism works technically....

Let us assume energy acting on a nuclide affects the entire structure together. This means energy going in or out of the system is treated as acting uniformly in time across the system. Aside from proportional distribution of energy, it will be as if the other groups aren't there.

Energy gets divided between angular rotation and linear expansion at contextual convenience. These act on the system as surface containers/manifolds: $U=(c\hbar/\delta Q)^2 \to 2\pi W_{kh}$. The electroweak spacetime (W_{kh} see pg. 140) consists of both intrinsic and transient (heat) values.

Our QM system is a quantum foam degenerate distribution. It undergoes an imaginary linear surface expansion. It is imaginary in that the change function is the sub-temporal distribution. If it were temporal, it could not function in this already occupied space. Ordinary solid geometry applies but under imaginary conditions.

		F	V=FS/J	E=F+V–2	S=2E/F	J=2E/V
Trionic Point*	–	1	2	1	2	1
Tetrahedron	s	4	4	6	3	3
Octahedron	p	8	6	12	3	4
Cube	d	6	8	12	4	3
Icosahedron	f	20	12	30	3	5
Dodecahedron	g	12	30	20	5	3

*renormalization of color charge shaping to spherical point.

10.4: Regular Solid Features

Mensuration formulas (below) for regular solids utilize Euler's and related variables (above). S=sides per face is easily computed by dividing F/360 then find the median of the listing of all the 3s, 4s, and 5s; or S=2E/F. J=number of edges joining a vertice, each edge joining two vertices.

---

[12] Kawanaka, H. et al. (2016). Enhancement of ferromagnetism by oxygen isotope substitution in strontium ruthenate $SrRuO_3$. nature.com/articles/srep35150.

[13] Nave, J. (2017). Bar Magnet and Solenoid. University of Georgia. http://hyperphysics.phy-astr.gsu.edu/hbase/magnetic/elemag.html.

Quantum Relativity

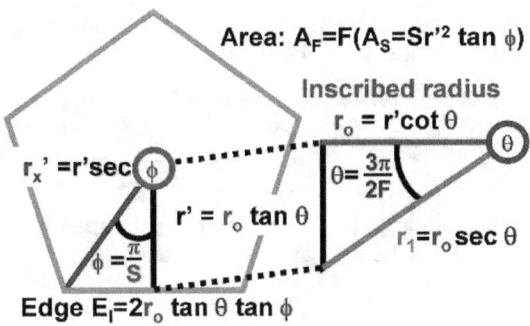

10.5: Mensuration of Regular Solids

Polyhedron evaluations renormalize into spheres but are enormously helpful in QM. Our focus is total area=A(F)=A. Each structure group/shell has its own proportion of the electroweak spacetime (manifold) of the whole: $u_A = U \, A/\sum A_n$. We used u_A here to avoid confusion with the angular U_A used in the density function.

Exscribed / Inscribed $\frac{r_x}{r_o}$ =		$\sqrt{\tan^2 \frac{3\pi}{2F} \sec^2 \frac{\pi}{S}}$
Tetrahedron	s	4.930 893 275 967 83
Octahedron	p	1.669 086 806 815 86
Cube	d	1.732 050 807 568 88
Icosahedron	f	1.109 302 142 000 00
Dodecahedron	g	1.123 450 055 497 63

10.6: Regular Solid Ex-Inscribed Radial Ratios

While treated as polyhedrons relative to each other (e.g. squeezing r_x of one into r_o of another), linear expansion smoothes into angular. To retain density and conserve value, the angular gets projected as magnetism. Squeezing consists of increasing the areas proportionally together until one can fit inside the other. Rescaling takes energy that could be magnetic force—inefficiency in the system. Radial ratios show p goes easily into d and f into g.

Electron Orbits

Atoms complete what to us would be the renormalization process. The components going into an atom are not alone completely renormalized. Our renormalized perspective is Wiswesser's system for electron energy sequence (below).[14] is It groups electrons according to Bohr's model

---

[14] Wiswesser, W.J.; Konier, D.A.; & Usdin, E. (June 30, 1972). <u>Wiswesser Line Notation: Simplified Techniques for Converting Chemical Structures to WLN</u>. Science: 176(4042):1437-9.

Quantum Relativity

(n=principle quantum number/energy level) and orbital angular momentum (l=azimuth quantum number=0→3).

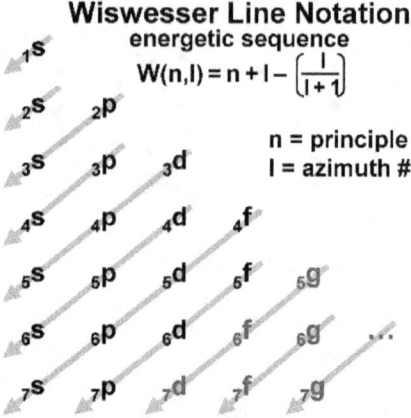

10.7: *Wiswesser Notation & Equation*

These quantum numbers are integers whose units are renormalization—confined quantum variables. Electron orbits satisfy two more: magnetic ($m_l=\pm 0 \rightarrow 3$ orbital positions in s=1,p=3|x,y,z,d=5,f=7) and spin ($m_s=\pm\frac{1}{2}$ parameter direction). Put together, the eigenfunctions provide the so-called "electron cloud" or Eigenstates maps.[15]

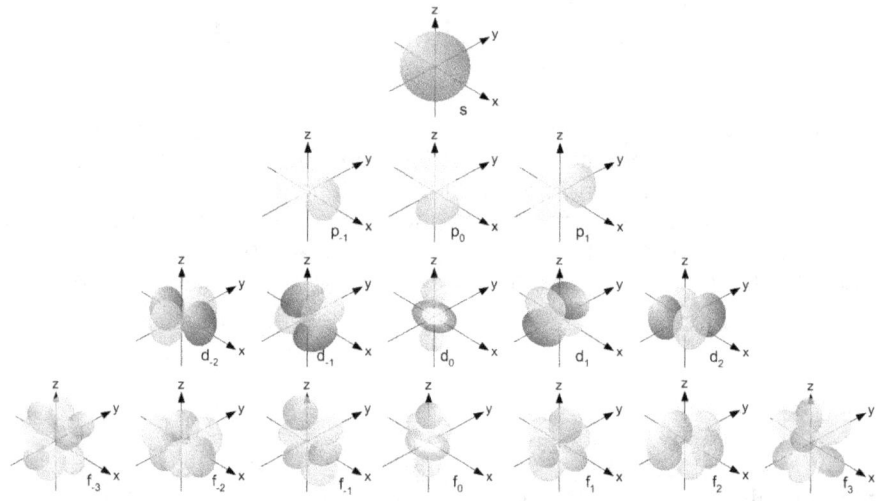

10.8: *Electron Orbit Eigenstates*

---

[15] Carolina, N. (Nov. 30, 2017). <u>5 Ways to Learn Orbitals in Chem 130 at University of Michigan</u>. oneclass.com/blog/university-of-michigan-ann-arbor/26649-5-ways-to-learn-orbitals-in-chem-130-at-university-of-michigan.en.html.

The problem with renormalization is confinement. Our outer perspective tells us most of how we interact and observe relativistically, but not everything. There is a reason for every physical attribute from thermal conductivity to the octet rule.

We need to differentiate our practical level thinking perspective from the confined quantum levels the qualities and rules emerge from. Electrons see themselves quite differently pre-renormalization. They see themselves the same way nuclide groups do: as equal regular lattices. Because they are quantum, they enjoy the QM aspect potentials (see pg. 153).

QM tends to be a rather broad category of thinking. Here we are taking it in its literal interpretation as mechanics of fundamentals. The s-orbit illustrated below has four vertices. We can think of this in semi-classical mechanics terms with a Heisenberg uncertainty contextual twist.

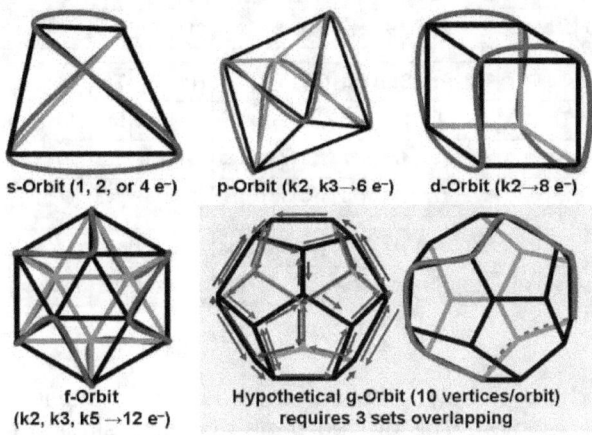

10.9: Electron Orbit Paths in Quantum Solids

Each electron consists of an entangled pair, dividing themselves across two vertices. Those same two vertices can be occupied by another electron by changing the spin quantum number. Another pair of electrons can be added at the remaining vertice pair by adapting azimuth to vertice.

This is like having four pit stops on a racetrack. Each electron occupies two while redefining its space in a fixed direction. There are two sides to the track, so two electrons can be at the same pit stops, and two more can be at the other two pit stops. Later orbitals often "drop" electrons down to this level and appear in two qualitatively different ways.

As a mechanical system, balance is required so you can have 1, 2, or 4 electrons in s, but not 3. Likewise, an orbit has to flow along one continuous line in one cycle without crossing over itself connecting all the vertices. The g-orbit is the only regular solid this cannot be done with. The best one continuous cycle can intersect on g is ten vertices. Additional paths cross over each other causing exclusion violations.

Traditional electron orbits fill in the Wiswesser sequence. Here they fill by vertice order limitations. This creates an entirely different set of possible configurations that can become incredibly ambiguous, but accurately

Quantum Relativity

predict valence states. It remains unclear, however, whether a g-orbit segment alone is possible, and to what extent it could apply.

Periodic Table

s																p	
H	He																
Li	Be				d							B	C	N	O	F	Ne
Na	Mg												Al	Si	P	S	Cl	Ar
K	Ca	Sc	Ti	V	Cr	Mn	Fe	Co	Ni	Cu	Zn	Ga	Ge	As	Se	Br	Kr
Rb	Sr	Y	Zr	Nb	Mo	Tc	Ru	Rh	Pd	Ag	Cd	In	Sn	Sb	Te	I	Xe
Cs	Ba	La	Hf	Ta	W	Re	Os	Ir	Pt	Au	Hg	Ti	Pb	Bi	Po	At	Rn
Fr	Ra	Ac	104	105	106	107	108	109	110	111	112	113	114	115	116	117	118
119	120	121	156	157	158	159	160	161	162	163	164	165	166	167	168	169	170
171	172	173	208	209	210	211	212	213	214	215	216	217	218	219	220	221	222

f:

Ce	Pr	Nd	Pm	Sm	Eu	Gd	Tb	Dy	Ho	Er	Tm	Tb	Lu
Th	Pa	U	Np	Pu	Am	Cm	Bk	Cf	Es	Fm	Md	No	Lr
122	123	124	125	126	127	128	129	130	131	132	133	134	135
174	175	176	177	178	179	180	181	182	183	184	185	186	187

g:

136	137	138	139	140	141	142	143	144	145	146	147	148	149	150	151	152	153	154	155
188	189	190	191	192	193	194	195	196	197	198	199	200	201	202	203	204	205	206	207

10.10: Periodic Table Limitations

If g could be entirely satisfied, we could also expand the possible periodic table (above). The grayed parts of the illustration require the g-orbit. Without the requisite environmental factors, related nuclides are unsustainable. Protons provide order in the nuclide, which requires the sustaining disorder of electrons. As such, complex nuclide stability depends partly on electron stability.

11. Degeneration

QM is mainly a predictive system for analyzing too little information. With atoms we finally achieved all our levels of renormalization. Quantum didn't go away just because we managed to achieve a relativistic universe in which classical actions are easily handled. It is everywhere in plain view.

Now we begin the processes of degeneration to whole new levels of quantum. Everything we do from this point forward is electromagnetism evolving back into fundamental forms. By electromagnetism we mean the range of fields with hyper-complex definitions. That awkwardly includes Einstein's surface gravity (the GFE), but not fundamental quantum gravity.

Renormalization and degeneracy provide the full range of our observational perspectives. Even light, a purely quantum phenomenon, is observed by its renormalized effects as virtual particles. Where renormalization enfolds the quantum, degeneracy unfolds a reflection. What is clearly angular momentum on one level evolves into magnetism and fractal-shaped filaments on others.

To a small degree we can explore degeneracy in the laboratory experimentally. Unfortunately, that small degree only covers our immediately practical. It doesn't explain the celestial and bigger pictures. For the greater things we can only observe the effect in incredibly slow motion, and experiment mathematically.

What is Degeneracy?

Degeneration is reduction of a set into arrangement-independent common variables; or a limitation that reclassifies to a simpler nature.[1]

Degeneracy is widely used and seldom defined, let alone defined clearly. Degeneracy describes complex systems of so many confined parts, or under other extreme conditions that their arrangements become cooperative as a quantum field or other collective phenomenon.

Any massive condensed population has collective if not QM qualities and behaviors. Molecules pooling into a drop is a sort of degeneracy. Arrangement makes no difference to identity as a drop. The drop is a completely separate nature from the molecules called a state (solid, liquid, gas). Large condensed populations of life, like nervous systems or local populations also develop collective and QM properties and behaviors.

A mole of hydrogen at 6.0221409×10^{23} atoms/g (one electron and one proton) is already degenerate—too numerous to count any way but statistically. It is even too numerous to count mechanically. Earth is

---

[1] Weisstein, E.W. (2018) <u>Degenerate</u>. From MathWorld--A Wolfram Web Resource. mathworld.wolfram.com/Degenerate.html.

estimated to be 5.972×10^{24} g,[2] estimating the number of protons and electrons at 7.19285×10^{48}. That is an inconceivable number of points whose arrangement is practically irrelevant.

The arrangement is basically fixed unit points in a field. Small variations in that set constitutes anomaly. Anomaly smooths out in the end to insignificant. Earth is a degenerate object. A galaxy filled with a hundred billion stars is also a degenerate object. The universe filled with hundreds of billions if not trillions of galaxies even on a relatively local level is also degenerate.

Matter degenerates under such extreme conditions that its properties and actions (individual and/**or** collective) become quantum mechanical.[3] This can include pressure-density conditions that exclude ordinary thermal pressure variables as with nuclides and degenerate gases (electrons).

Degenerate electron orbits are arrangements with the same energy (pattern)[4] as with white dwarfs. Degeneracy is often associated with fermions and the Pauli Exclusion Principle.

Exclusion forces electrons to differentiate into a "degenerate gas" subject to Heisenberg uncertainty.[5] Instead of having mechanically fixed positions in space, the parts are dynamically quantum. And, instead of getting larger with mass, they simply grow more dense.[6]

We have an incredibly broad range of potential applications for degeneracy. Once atoms have confined fundamental quantum mechanics, the whole process reverses itself to re-establish new levels and return to the simplicity of fundamental. It doesn't wait for another generation of matter. It begins even while renormalization is buttoning the last confinements down into atoms.

A critical feature of degeneracy is to disconnect: first of relative from quantum, then classical from relative. Quantum phenomena don't go away with these disconnects, but our ability to make sense of them goes awry. We can see at the quantum levels how energy enters or otherwise affects value in an identity. That value awkwardly translates into a change in momentum.

We say awkwardly because there are always far more variables than we can reasonably observe or account for. Some have directional influence, while others just passively increase magnitude. In each case, the effects get washed out together in relativistic momentum. The reality,

---

[2] Cain, F. (Dec. 24, 2015). <u>Earth's Mass</u>. universetoday.com/47217/earths-mass/.
[3] Taylor, D. (June 2012). <u>The Life And Death Of Stars</u>. NW University. faculty.wcas.northwestern.edu/~infocom/The%20Website/pressure.html.
[4] Hunt, I. & Spinney, R. (2006). <u>Organic Chemistry</u>. McGraw Hill.
[5] Rieke, M. (Mar. 18, 2002). <u>Lecture 17: Stellar Evolution</u>. ircamera.as.arizona.edu/astr_250/Lectures/Lecture_17.htm.
[6] Buckley, J.W. (Mar. 7, 2013). <u>Degenerate Equations of State & White Dwarfs</u>. St. Luis: WA University. physics.wustl.edu/buckley/556/Theory/degeneracy_jwc.pdf.

however, is a complex of changes from as intimate as microstates to as overt as expansion and bounce.

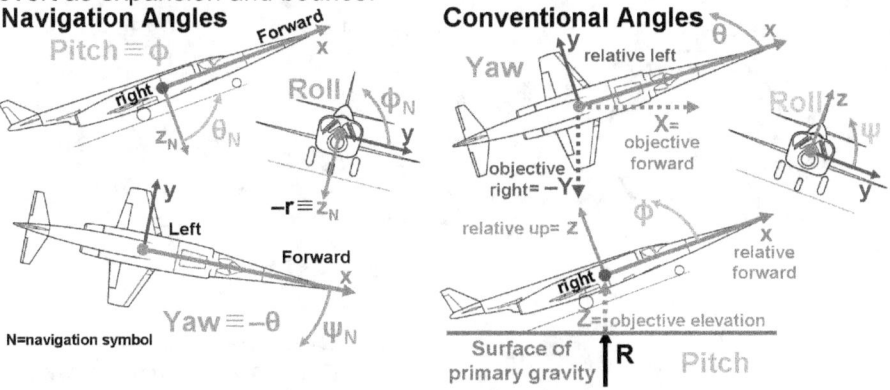

11.1: Classical Navigation Axes and Angles

We want and expect things to work as easily as navigational axes and directions. We can't even find agreement between our own simplistic systems, or relative perspectives. In each of these diagrams, the upper case letter is the intrinsic axis and lower case is the directional axis.

Above, forward is always x and left is always y, without the ambiguity of what exactly constitutes front. Convention follows Euler's spatial construction. Navigation's Tait-Bryan (Cardan) angles[7] are derivative and inconsistent. Whether you are a physicist or engineer, the practicality of air traffic control is a wake-up call.

It is bad enough that variables get recycled and layers of ambiguity are added. Being inconsistent with the variables is seriously problematic. Variable identifiers are no less important than names and unambiguous definition (e.g. image below). Failure to be consistent can be catastrophic. Of course the consequences are less obvious in physics theory.

Navigation puts direction (xyz) relative to the object in motion (XYZ).[8] Conventional analysis, however, uses the objective environment to define where the object is (XYZ). This affects perspective of the directional axes (xyz). From the conventional perspective, the Earth is so large that the local surface is an equivalent XY plane. Z is then the elevation above said plane. Gravity-wise, negative Z would be radial R beneath the confining mantle.

We get no good answer about up, down, front, and back until there is enough quantity that an entirely new set of features emerge—another level or sub-level of degeneracy. Just as we have difficulty weaseling out the differences, so too does the universe. Unlike us, the universe isn't sitting around debating about it. It simply follows the path of mindlessly least

---

[7] Abraham, S. (2017). Rotations and Orientation. cs.utexas.edu/~theshark/courses/cs354/lectures/cs354-14.pdf.
[8] Hall, N. (May 5, 2015). Aircraft Rotations. grc.nasa.gov/WWW/K-12/airplane/rotations.html.

resistance. We can consider the same systems adapted for change or equivalent quaternion axis functions.

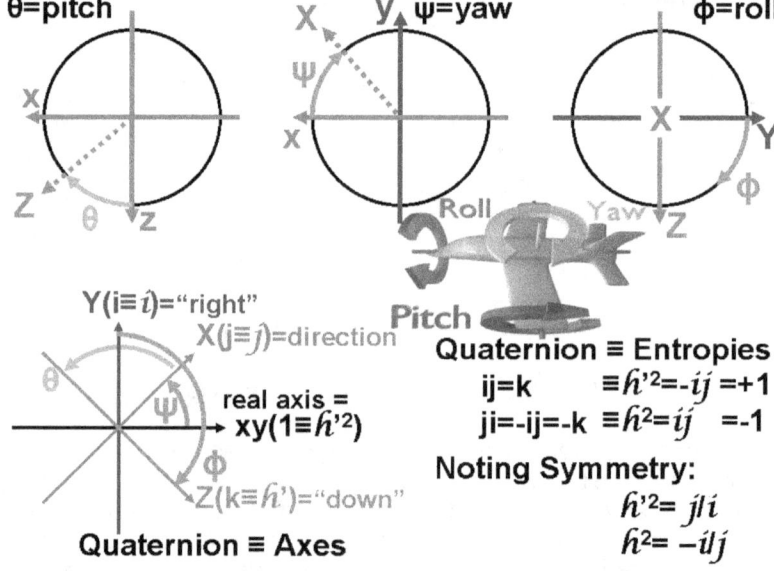

11.2: Navigation by Quaternion/Change Axes

Setting aside symbol conflicts (e.g. psi and wave function), even a simple photon borders on unmanageable by hand computations. Evaluating such a data set is evaluating nature's analog to binary machine language of a program instead of mechanically processing it. In other words, Bell's Theorem applies: too many variables, and hidden variables, requiring quantum mechanical means to adequately process.

As a sub-routine within an object oriented algorithm structure, however, we can ignore local symbolism conflicts in other routines. This makes computer modeling particle behaviors as accessible as plugging in an established data set. The universe handles the data set exactly as presented to include all distractions in their degrees. Start simple and use that to better analyze complex data sets. As interactions merge in degrees, these sub-routines also merge in degrees up to disconnect.

Degeneracy confines complications in extreme conditions such as extreme density or quantity. This confinement on one level simplifies, and on another disconnects leaving the mystery of hidden variables. Degeneracy is generally described as a condition with quantum behaviors, but what exactly are these quantum behaviors? Some of them are as ordinary, classical, or relativistic as thermal expansion or electrical current. Everything in the quantum universe is quantum field derived.

Beyond Wonderland

Neutron and electron degeneracy pressures are the two faces of the wizard of Quantum Wonderland. From them follows uncountable degeneracy—so many things you can only count statistically. The numbers like Avogadro's 6.0221409×10^{23} atoms/mole (protons/gram) are so incomprehensibly large they can't even be counted mechanically.

We live in this intermediate uncountable dust of the Earth level of degeneracy. Below us is the uncountable number of atoms per gram. Around and above us are the uncountable grams making up our world. Our uncountable dust reality is Wonderland.

Uncountable degeneracy is the nature of celestial objects. The next level is countable, assuming you can find all the interacting celestials making up a system. Next: back to nearly uncountable solar systems in a galaxy, itself in a countable group of interactions in an un-findable and uncountable number of interactions making up the universe.

To properly classify matter poses challenges. We are classifying by the context of interactions. Particle classes are fairly easy to follow through confinement, renormalization, up to degeneracy pressure where the whole paradigm flips over. The focus switches from infinitely divisible energy to whole number units of matter.

It is little wonder that we want to ignore the conveyance of value by light and think exclusively of things in terms of material conveyances. Our Wonderland is in the middle looking forward and up-ish. Below the numbers are too many, and the infinitely divisible unfathomable. It is at least equally relevant. We just don't give it that much respect. It is far easier looking at and working with mechanical clumpings.

While we focus on these irrelevant details, the universe prefers fields. A field defines all its points no matter the scale given to those points or system of mapping them. Fields are conceptual abstractions easiest to represent mathematically. Here is where the concepts of degeneracy get confusing. At some point a field boundary condition gets triggered creating an identity that looks like a primitive identity but at a completely different scale. The first of these are celestial WIMPs.

Celestial WIMPs are hypothetical weakly interacting massive particles. Traditionally these are sought in particle accelerators—the physicist's version of a microscope. Reality is that they are dark matter large enough to act as nucleation points to form celestial objects and celestial object - level magnetic fields. They are super-sized sixth generation matter (e.g. degenerate pressure identities). We see evidence of them in magnetic fields and by their hexagonal shadows at the poles of some worlds.

Earth's polar hexagons are rough due to resistance of solids. Saturn's (below) are the most prominent but smoothed by gas motion.[9] Neptune's are hard to see due to smoothing and coloring. It is hard to gauge the size

---

[9] Stricker, M. (Jan. 14, 2017). The Storm System at Saturn's North Pole. www.sun.org/images/the-storm-system-at-saturn-s-north-pole.

of the oblation or prove it is a WIMP shadow. Very likely it is a degenerate particle the size of a house or even bigger.

11.3: Saturn's North Pole

Clearly an object with such a magnitude of interactive potential is a tad beyond human ability to emulate. And there it is. Reality check.

A singularity can occur at any magnitude. That does not mean the singularity can be maintained or grow from any magnitude. Looking at weak bosons, we see order and disorder systems of one magnitude. WIMPs are a hypothetical greater magnitude of the same kind of quantum phenomenon. The overlapping bubble system of singularity-horizons is another weakly interacting system.

Among the interacting bubbles, the same values are defining volume and surface simultaneously. That satisfies the GFE—the mass omega variable. But then there are charge interactions creating massive electromagnetic field lines. Both are reasonable contributors to the clusters and filaments of galaxies we observe, and their interactions further contributing to the omega factors compacting the ideal spacetime horizon to a humble 46.85 Gly.

We can begin to emulate what is above, by working with what is below on condition that we recognize the differences due to boundary conditions. Singularity behavior on neutron star magnitude interacts with clouds of atoms. It can rob a neighbor star and even feed the atoms back.

A singularity or neutron star of stellar magnitude cannot consume a whole star at once. Only in pieces. It can only consume a lesser magnitude of disorder. If it is a singularity, it can only consume matter. It cannot be light because light cannot interact directly with a singularity. Classifying these objects depends on the level of their ability to interact. The limitations of interaction further affect our ability to accurately reconstruct in physical laboratory conditions. We can, however, computer model the fields applying the rules and boundary conditions.

Bell's theorem basically says that we have to use Quantum Mechanics because there are hidden variables we may never know. Even when we do know the variables, we cannot collect all their information let alone have the computational capacity to evaluate that information.

We can use our understanding of hidden variables to better model and simulate the Quantum Mechanics we cannot observe to improve our ability to predict. We may ultimately know many things, but it is not the job of science to know. It is the job of science to explore in the pursuit of understanding. There will always be more.

General EM Fields

The trick of experimental mathematics is proof: arriving at established solutions, observations, structure, or other logics. The hazard, of course, is that mathematics can provide a never-ending flow of problems going nowhere. Hidden quantum variables add injury layers to this hazard. It is thus imperative to work our variables back to easily quantifiable measures or verifiable established functions—classical proof to axiom.

The Matrix we introduced earlier (pg. 97 et seq.) remains an intriguing mathematical curiosity. We continue to study it for patterns linking it to working constructs. We may not be clever enough at present to distill a perfect way to explain it, but it is helpful on several levels. The Matrix remains as much a quantum mystery as the universe it helps us study.

Electroweak: $\qquad w_{kh} = i(r_k \partial_h - r_h \partial_k)$

$$\frac{r's + rs'}{ij = 2\hbar^2 \to t^2} \Rightarrow \frac{\partial_p \sqrt{r'^2 + r^2}}{ij\rho} \Rightarrow \frac{\partial_p (r' + i'r)}{\Delta}$$

$$\Rightarrow \frac{-i\partial_p (r' + i'r)}{i'\Delta} \Rightarrow \frac{i\partial_p (i'r' - r)}{\Delta} = e_w$$

11.4: Electroweak Field Equations

Aside from a breakdown of fundamental variables, the Matrix gives relationship clues. Of particular importance is a significant difference in variable perspectives. The spin forces (s' and s), as an example, are a means to convert linear to angular forces. They also pave a convenient path to Laplacian spacetime distribution. Consider the electroweak spacetime manifold (w_{kh}) above.

We noted earlier $r_k = g_s^{-1} = \nabla$ (where $\nabla^2 = \Delta$) links to Gauss: $g_t/g_s = 4\pi G \rho_m$ (dilation=g_t=/s² and mass density=ρ_m). Where $\rho = (s's = c^2) = \Delta_t$ is a temporal Laplacian distribution operator, and $ij\, \Delta_t = \Delta$. We derive to extract the spin variables as distribution. On the Matrix, weak is e_w. As gravity and thermal at the same magnitude interacting, it comes first in the EM hierarchy.

Disorder (i and i') as change operators inject directly into the Laplacian distribution and ordered/intrinsic linear force (r'). The ambiguities of form in the variables allow this to be evaluated in a variety of units, or as unitless scalars into quasi-temporal spacetime manifold $t^2/m^2 = w_{kh}$.

The most intriguing part of this emerged when we realized how this applies to nuclides. We assumed permittivity meant the entire space was occupied period. This showed functional exception that not only didn't blast it apart, but contributes to preventing the nuclide from enfolding. The quantum universe is inconveniently fluid and malleable. Degeneracy is at least as ambiguous and nuanced as our attempts to define it.

Magnetism is both angular and spin force degeneration creating a combined sub-temporal deformation in a spacetime distribution. Where the other variables localize, magnetism puts value at a distance freeing local resources. Its elements account for multiple post-orbital EM definitions.

Locality of value for is of particular significance if you are a star. Some of your energy is better being projected proportionally away. Creating a lot of highly energized ferromagnetic elements makes "away" in the wrong place. Even though the elements are a lesser magnitude, too many will bump the energy production to the wrong magnitude in close range.

The Matrix provides two possible Laplacian distributions for each linear and angular evaluation. These often apply together as is the case with magnetism.

$$\Delta_{ht} = z^2 = xy = \sqrt{x^2+y^2}$$
$$\text{and}$$
$$\Delta_{xt} = r^2 = r'r = \sqrt{r'^2+r^2}$$

These and many other Matrix-generated forms remain largely unexplored, but appear to provide insights into topics like linear thermal expansion. The problem with these degenerate pursuits is the sheer numbers and layers in each of those items to explore.

Operator-values z and r are also forces and, $rz = \rho = \Delta_t$. In their active roles, as operators they create directional charge fields: z (angular and polarized) and r (linear). Magnetic spacetime distribution is:

Vector B to free energy $A\nabla =$ $\qquad B = \dfrac{\partial_p z}{\nabla}$

$$\dfrac{xs + ys'}{ij = 2\hbar^2 \to t^2} \Rightarrow \dfrac{\partial_p z}{ij(\rho = \nabla_t^2)} \Rightarrow \dfrac{\partial_p z}{\nabla^2} = A$$

11.5: Magnetic B Vector and Free Energy Fields

The phase operator ($\hbar$) provides the characteristic toroid shape. There is no difference swapping ∂_x for ∂_p. There shouldn't be because we are simply swapping which variables are providing value versus function, and these forces are inter-dependent. Of course extreme conditions will completely disagree with indiscriminate swapping. Like other EM fields, we can expect this has a range of evolving applications differing across a spectrum. Magnetism is an evolution of angular momentum shifting object orientation from around (circular) to adjacent (toroidal) and then evolves to arcs into beyond.

$A=B/\nabla$ is the Helmholtz "free energy" ($A \equiv U-TS$) in decomposition of magnetic vector potential. The magnetic vector field (B) is a sub-temporal force direction. Diverging $B \nabla \rightarrow 0$ gives the force no space to function in, neutralizing field direction per Gauss' Law[10] as with electroweak $w_{kh} \rightarrow c^{-2}$ Both already describe divergences as containers (manifolds). Putting divergence (c^2) in that container cancels the container.

The spin elements of electroweak (linear) and magnetic (angular) interactions make them simpler. They still experience osculation— perturbations caused by interactive field detail differences including n-body issues and aspect perspectives.

In electromagnetic interactions, values are shared. The local observer may feel like the values are being conveyed, but the conveyance is enclosed/contained by the system of interaction. You are subject to the force of surface gravity, and though it affects you in degrees relative to surface anomaly, it does not convey. It is shared.

This is a massive clue to how a star can shine so bright and gain in mass rather than lose it to emission. It isn't emitting, it is defining spacetime. Just as a singularity is a focal point of enfolding, a star is a focal point of unfolding. It is ultimately reducing/simplifying to expansion or contraction of spacetime. Nothing actually needs to be lost, only converted.

In our intermediary position we perceive transfer of energy acting on, creating or evolving matter. We are observing local perturbations and their interactions. On a grand scale, however, the scalar forces are just spatial values without direction to specify their own spaces. They don't own spaces. Spaces own or at least temporarily borrow them.

Dark energy and dark matter aren't just expanding void and singularities, but also stars and anomalous contraction or bounce-back clouds in spacetime. Einstein argued: space does things. Let us add it does them in quantum ways, and that is where things get confused. Einstein's relativistic static universe does nothing in the most incredibly quantum dynamic way.

These fields are spacetime manifolds constructed the same way as Einstein's. All we did was expose the fundamental quantum elements. The EFE are mathematically degenerate if for no other reason than the variables convey into a different (albeit related) class. To be real each variable is also hypercomplex. The synonymous natures of mathematical and physics language and rules should never be ignored.

Attention span and individual wants top our classification problem. It is hard enough to get the word VIRTUAL to sink in with photons and light. Add to this other expectations and wants and calling the EFE a form of EM goes sideways. They are as alike as $2x+3y=z$ and $5a+3b=c$. The Matrix

---

[10] Horsley, R. (Mar. 26, 2014). "9.5 Helmholtz's Theorem" in <u>Hamiltonian Dynamics</u>. University of Edinburgh. www2.ph.ed.ac.uk/~rhorsley/SI09-10_t+f/lec19.pdf.

makes distinctions, but even here we see too many different variables using the same symbols.

There is nothing electric, electron, or magnetic about gravity. They have things in common, like how their fields get defined mathematically. They mostly have very different variables, but without the order of gravity you can't really have anything else.

Orbital "planes" confuse just about everything. They contain linear and angular exchanges, giving them a range of reference and change shape potentials that quickly enfold into operators. The math simplifies into conic sections and circular functions that show some of how our complex forms get applied.

$$\frac{r'y + rx}{ij} \Rightarrow \frac{\partial_z r}{\nabla^2} = e_o$$

Also true: $\partial_r z/\nabla^2$, or more ambiguous as noted earlier. Newton's position ($\rho = re_r$) uses linear $r^2 = x^2 + y^2 \equiv r'^2 + r^2$. The change information gives angle $\tan\theta = x/y$. The ij-change is $e_r = (\cos\theta, \sin\theta)$[11] leading to velocity $v = r'e_r + re_r'$, etc. The ambiguities are linguistic attempts to simplify on one hand and give better meaning on the other.

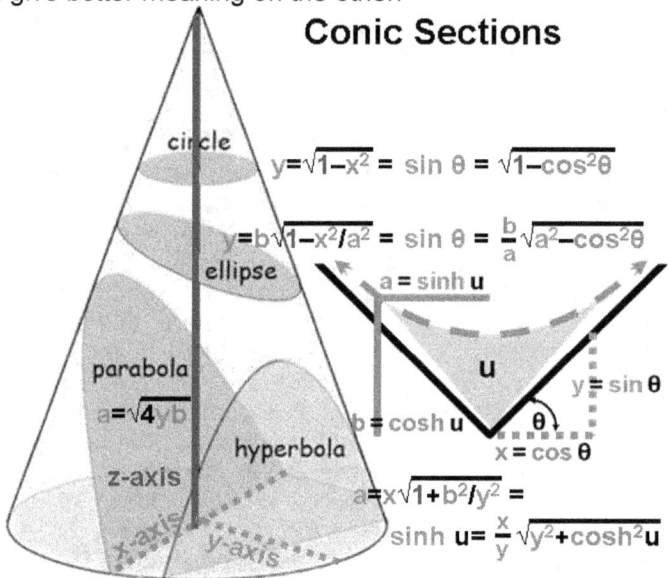

11.6: *Conic Section Recti-Polar Definitions*

---

[11] Fitzpatrick, R. (Apr. 26, 2011). Newtonian Dynamics. Austin: University of Texas. p. 57 et seq. farside.ph.utexas.edu/teaching/336k/Newton.pdf. Miller, W. (2006). 0.1 Newton's equation with gravitational force. https://www.ima.umn.edu/~miller/1572Newtongrav.pdf.

Angular momentum (z) is traditionally given unit value and ignored. Treated as derivative of linear, angular change is applied as $e_z = (-\sin\theta, \cos\theta)$. We know it isn't so simple. Changes in any one value will make an orbit stretch, contract, or oscillate.

An orbit can tend toward linear or angular affected by imbalances. Exaggeration of one angular element can lead to a sinusoidal feature and exaggerating aspect oscillation into a twisting rotation. On primitive levels, the variables reduce to one linear, one angular. As degeneration evolves in the generations, the variables diverge (like twists in spiral arms) and eventually invert (e.g. filaments).

As an interaction, each body presents its own orbital plane perspective. Generally, orbits increase order, such that the Kepler orbit interaction conveys the potential for orbital twist to polar oscillations and the orbital azimuth. Adding bodies to a problem increases disorder.

Increase in disorder increases the wave pattern of the plane. This is notably evident in lunar month anomalies,[12] Mercury,[13] and asteroids. This perturbation is called orbital **osculation**. Osculation details provide exact orbital description but are computationally laborious, as opposed to mean orbit generalities.[14]

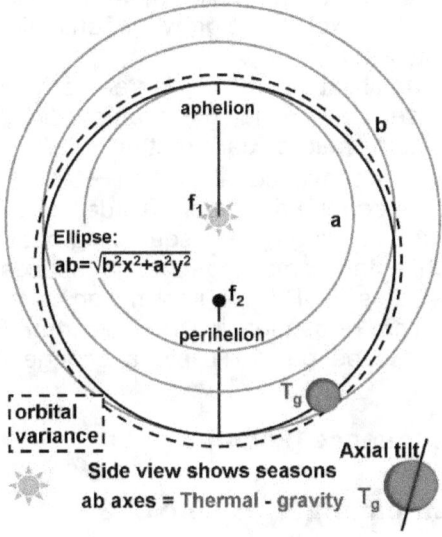

11.7: Basic Elliptical Orbit

Flat representations of orbit, like the one above, don't do the full range of features any justice. Oscillation and other complexities aren't as easy to

---

[12] Espenak, F. (Jan. 12, 2012). Eclipses and the Moon's Orbit. eclipse.gsfc.nasa.gov/SEhelp/moonorbit.html.

[13] Balogh, A. & Giampieri, G. (Mar. 20, 2002). Mercury: the planet and its orbit. www.astro.wisc.edu/~ewilcots/courses/astro340s04/readings/mercury.pdf.

[14] Guinn, J.R. (Dec. 28, 1995). OSMEAN-Osculating/Mean Classical Orbit Elements Conversion. ntrs.nasa.gov/search.jsp?R=19940003221.

imagine as tilt in the plane and elasticity. It also doesn't illustrate this is the byproduct of two fields interacting into a common system.

Unfortunately, we did not invest time into orbital research. We will go no further on it. Our point is the ambiguity of forms for modeling further study. One electromagnetic function can be interpreted in a broad array of ways based on developmental hierarchies, value distributions, etc. One has to be careful with these generalizations to account for context, let observations have the last word on interpretations, and first word in theory.

Degenerate Matter

Singularity is ultimate order degenerating to fully occupy all or part of a j-entropy spacetime. Primordials only define a part of a j-entropy. This perfect distribution of value is the ideal quantum singularity. Alone, it annihilates itself because there is nothing to stop it from enfolding.

Nucleons and neutron stars are degenerate and singularity-like with an intrinsic level of disorder preventing annihilation. Unlike quantum singularities, these fields are relativistic: more intrinsic than environment-defined. We now need to explain the growth of enfolding and the degree to which unfolding is localized.

What prevents annihilation is the degree to which unfolding value (divergence) is localized and enfolding (decomposition) is diminished. This optimizes at the $v=1:\mu=e$ ratio of dark matter (1–e/(1+e)=26.894≈27%) to dark energy (~68%) plus conventional matter (+~5%).[15]

Wheeler called the degenerate order-disorder mixture quantum foam.[16] It is like a solid sponge soaking up and squeezing out value. One quantum number is filled while the remaining change spaces remain available. Sequence (k) constitutes EMR:EMA information, aka foam, the specific distribution details. This re-sequencing derives from Schrödinger's wave normalization function (see pg. 131) into a change distribution form of Laplacian.

Change Sequence (Foam) $\quad k^2 = \dfrac{\partial^2 R}{\partial R^2} + \dfrac{\partial^2 A}{\partial A^2}$

Foam Distribution $\quad \partial_i \varepsilon + \partial_j \varepsilon = \nabla_{k'}^2 \rightarrow c_{k'}^2$

11.8: Strong Type VII—Foam Redefinition

Change in k-information=k' includes information equilibrium and evolving toward uniform change distribution (all positive with positive, all negative with negative). This runs on the cosmic clock independent of

---

[15] Nagaraja, M.P. (Ret. Feb. 28, 2018). Dark Energy, Dark Matter. science.nasa.gov/astrophysics/focus-areas/what-is-dark-energy.
[16] Wheeler, J.A. & Ford, K.W. (2010). Geons, black holes, and quantum foam : a life in physics. New York: W.W. Norton & Company.

everything else. Foam distribution is the phase tangent (derived instant rate of change) of value functions in a complex spacetime (c_k).

Information changes occur in temporal or quasi-temporal perspective with changes in value roles (permittivity and permeability; order and disorder) and their distribution. That distribution contains internal and external perspectives. This is Thermodynamic. It will naturally follow the path of least resistance of order to disorder sequencing.

One way to look at this is by how an object defines its space. An electron defines its space by relative permeability ($\acute{\epsilon}$). A proton defines its space by relative permittivity (ϵ), even though we simplified the function to permeability. Adding mass to any object increases its use of volume. The effect has counter-intuitive consequences for leptons.

Electron degeneracy is due to leptons primarily using area. Adding mass to a lepton reduces their use of area by defining more ordered volume toward permittivity. This is observed with mass increase with radius decrease in white dwarfs.[17] Of course this caps out when the intrinsic volume gets in the way by filling its quantum number.

The opposite is true for objects with permittivity focus like nucleons, nuclides, and neutron stars. They politely increase volume with mass toward permeability (e.g. G) but retain their degenerate density. This accounts for radial change but not emission or interaction.

Interaction is a function of information equilibrium. To become part of a primordial degenerate (singularity) requires having the same simple information distribution. That is why role change is vital. This is yet another class of strong interaction (**type VII=strong equilibrium**=e_S). Since it is a foam function, let us call this anagenetic process the **Wheeler interaction**.

Through this process, some input is potentially hijacked by the identity. This "growth" can trigger decay or other form of degenerate evolution. In the process, new identities are formed, violate exclusion, and are ejected. Excess also "bleeds" out. Bleeding can be fairly uniform across a surface assuming an anomalous rather than ideal singularity surface.

We are in observed but otherwise unexplored terrains. It is reasonable to speculate within one degree of empirical information. After that, even the best hypothesis is roughly equal to pure fantasy. There is very much more we can hypothesize about from here, but too little information to even properly model. That isn't to say we are completely helpless.

Understanding the Matrix

The most important thing to learn is how you learn. That helps solve a lot of observer biases and effects. It also prevents you from being in a perpetual spin cycle. Repeating the same behavior reveals nuances and physically programs the mind to that behavior. It does not provide understanding, only perspective. To gain understanding requires diversity

---

[17] Rieke, G.H. (Nov. 24, 2007). White Dwarf. Tucson: University of Arizona. ircamera.as.arizona.edu/NatSci102/NatSci/lectures/whitedwrf.htm.

Quirky QR

and failure. That means not taking Feynman's perspective that if it isn't on his microscope slide it is pseudoscience. Physics is the bedrock of science. Everything else derives from and adds to those rules.

Diverse pursuits in physics are equivalently rigid and real. So too are other academic pursuits. Any field is only as much as you put into it. Everyone starts at the bottom in each field. There are no shortcuts in learning, but mastery and expertise in one can help accelerate or blow your legs off. Einstein slipped in Russell's epistemology. Being a world class of any one thing does not mean world class of anything else. Each stripe is earned from scratch individually.

We mention this because the Matrix was once considered a mathematically curious failure. Then one day it was observed emerging unexpectedly from a structural analysis of developmental psychology. The break in inhibitions opened lines of communication between previously disconnected neural segments. These aha moments are instances of understanding that connect previously disconnected things. You can only do that by actively working the disconnected things in depth.

So now we had to take the Matrix more seriously. Unfortunately, like the social sciences, it is non-linear. Not just superficially. Practically. You can derive a form from it, but until you have gotten your hands dirty with it, you cannot be sure how that form actually works or when.

As you get your hands dirtier, you realize one derivative forms a whole class of forms like the quadratic form: $\sqrt{i^2+j^2}=2(ij=\hbar^2\equiv t^2) \rightarrow \partial i+\partial j=dt$. Not to mention radian and other roles. Mathematical reasoning just got exponentially harder thanks to quantum.

The Matrix is three-dimensional, giving each variable three coordinates: row, column, and sequence. Each variable is an archetype within a group. Gravity (g) is a linear contracting spacetime. It appears as a momentum vector (kg m/s) with other vector-like archetypes: thermal (T), centrifugal (C'), centripetal (C), and EM (e). We should note that the Matrix provides at least a dozen forms of EM (e.g. hypercomplex).

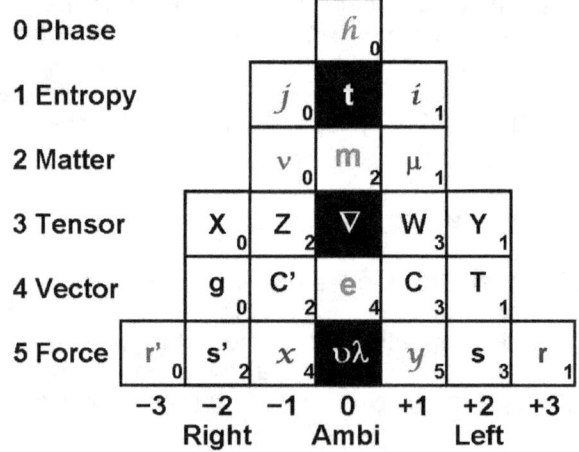

11.9: Periodic Matrix Surface Features

Quantum Relativity

All but three spaces in this are intrinsic. The black background dimensions are distributing functions (time=t, operator=∇, light and spacetime constraints c=υλ=∇/t). As distributing functions, they can be intrinsic OR extrinsic. As distributions, they are antithetical to and restrict their intrinsic cousins. These restrictions provide a sequence rate so everything does not happen simultaneously.

To evaluate relationships among the variables, we create an exponential mock-up. This can be written more simply as coordinates (r,c,s) or as a 1x3 matrix recognizing that multiplication adds the coordinates, division subtracts.

$$c^2 = WXYZ \div ij \to t^2$$

$$\begin{pmatrix} 4*3 & =12 \\ 2-2+1-1 & =0 \\ 0+1+2+3 & =6 \end{pmatrix} \div \begin{pmatrix} 2*1 & =2 \\ 1-1 & =0 \\ 0+1 & =1 \end{pmatrix} = \begin{pmatrix} 12-2 & =10 \\ 0-0 & =0 \\ 6-1 & =5 \end{pmatrix}$$

$$m \quad c^2 = \quad E \quad = e\sqrt{gC'CT}$$

$$\begin{pmatrix} 2 \\ 0 \\ 2 \end{pmatrix} \begin{pmatrix} 10 \\ 0 \\ 5 \end{pmatrix} = \begin{pmatrix} 12 \\ 0 \\ 7 \end{pmatrix} = \begin{pmatrix} 4+(4*4)/2 & =12 \\ 0+(2-2+1-1)/2 & =0 \\ 4+(0+1+2+3)/2 & =7 \end{pmatrix}$$

11.10: $E=mc^2$ Matrice Form

Einstein's $E=mc^2$ consists of mostly confined variables. The vector forms, for example, come together generically in E. That form can be written at least a dozen other ways just by analyzing the parts and understanding some of the other forms. Likewise with $WXYZ/ij=c^2$.

The first thing we realized is that the absence of one variable in a group (zero value) has no bearing on group existence. An identity always consists of all its parts. If you take away one of those parts, it is a different identity or identity is lost. Relativistic identities, however, are group proportions. Taking an item out of a group certainly has effect, like creating a potential, and requires that gap be filled some way somehow.

The second thing we realized was that the manifolds don't act like we expected spatial dimensions to act. This started with recognizing we could plot four spatial dimensions relative to cubic vertices. We quickly realized that our convenience probably has no use whatsoever in the real universe.

Of major significance are manifolds from lesser orders to the left contributing, and outgoing manifolds to the right. It is imperative to follow the natural evolutions of manifolds through the generations to appreciate the emergence of more complex fields and their shapes.

It took an incredible amount of time to realize the generations of matter or how complex operators fit in. John Dewey observed compounding in the cycle (applied to epistemology). We realized there was a tad more to it than that. We were very grateful to see biology has an exact analog to the system of generations we devised. It also clarified a few concepts like individuality and gave us another very useful layer of grouping (virtual, confining, system).

An incredible amount of time was nearly wasted examining geometric properties and potentials. Even mistakes are to be learned from. What we realized going through that was an incongruity that allowed a function to go one way but not the other.

Sequence was the last quality added to the Matrix, which changed the shape of the Matrix. It gave a level of numeric analysis in which we could identify the left-right (Fleming) properties and generalize scales. The left-handed dimensions all have odd sequence numbers. The $e^{3s}=(Uh)^{-2}$ modifier evaluates contextually (U) to e^s = 1.3E22 seconds=417 trillion years—just short of the limit to spectrum frequency dilation.

If the Matrix does nothing else significant, it gives us a way to keep track of our variables. While entertaining to study, and occasionally a great source of inspiration and ideas, we do not advise relying on it because it is a trickster. One must explore the Matrix with empirical caution and humility.

Conclusion

Most of our USE mode experience of the universe is nested in part of early stage degeneracy. We can look through our telescopes, send satellites and even craft to explore, observe, and document an ever-complicating universe. The sheer numbers of these complications ARE degeneracy. The anomaly of individuality is as significant as the differences between one grain of sand to others and the beach in total. Our world in this is a grain of sand.

Einstein's Relativity gives each body its own relative temporal position and rate. As we said earlier, the quantum universe is as ordinary as our social universe. Each individual is on their own path, at their own pace. Each was born at their own distinct time, will die at their own distinct time. Many have come and gone. Many are still to come and go. Each is an evolutionary potential that can also dead end.

Our social universe is a multiverse of individuals. Our physical universe is a multiverse of galaxies. Like our social universe, these individuals are born, evolve, interact, and die at their own paces, their own directions, in their own timeframes. Big bangs are all over the place, and galaxies are at various stages of development and observability all over the place.

This temporal dynamic is a massive monkey wrench in redistribution of expansion and value in void and CMBR. There is no cosmic NOW per se, only local relative NOW. As compounded fields, our observation of CMBR at a point here is not the same necessarily as it is somewhere else. We are at different points of time and dilation. What seemed simple is instead impossibly ambiguous.

The universe rhymes and even homonyms, but sometimes things just ARE with no particular cause like two cars colliding in an intersection. Each has its own cause for being there, but a collective cause is circumstantial. Circumstantial means an opportunity exists, like the road, weather, or vehicles. A circumstance can intercede and be causal by interaction. A circumstance can just be a passive element simply providing the setting.

We need to be careful with attribution of significance. Observing an interaction need not be made any more significant than the observation. The observation is significant enough. We don't need to go fifty steps removed to something else. It may not give great business results right now, but increased understanding evolves so the business doesn't die out by being stagnant, or by misinformation misguiding.

A black hole, the most powerful thing in the universe, cannot remain static. It has to grow and evolve its environment or it dies. This is true at all levels of existence. Biology has its own rules, but those rules are first limited by the physical universe. Social dynamics has its own rules, first limited by the bio-ecological universe. The interactive rule trickles down to every level.

By giving any one position too much significance, we miss the others. We may acquire an incredible knowledge but completely miss out on any understanding. We also set ourselves up for sustainability failure and lose integrity. Integrity loss damages science, invites crackpots, and undermines the bedrock upon which civilization plants its foundation.

Science needs to remember what is taught to our children. It is okay not to know. It is not okay to cling to ignorance. That is static. Break in the need to adapt, evolve, and the value of failure to learning.

Look things up. Actively engage research, not passively read the daily headlines. Endeavor to understand. To retain our integrity, science must accept that it does not know. It is the job of religion to know. We are allowed to say, "I don't know, but I'm working on understanding it." It is healthy to ask questions and to investigate. It is unhealthy to evade falsifiability and shape reality around speculation.

When your observations disagree with your theory, you have an obligation as a scientist to figure out why. Theory is always on trial. New evidence is not there to prove it via confirmation bias. It should be there to challenge it. Theory must always be ready to evolve or be replaced with better. Theory must always be built upon the rules established through empirical evidence.

www.ingramcontent.com/pod-product-compliance
Lightning Source LLC
Chambersburg PA
CBHW071053240526
45471CB00015B/1711